TRANSACTIONS

OF THE

AMERICAN PHILOSOPHICAL SOCIETY,

HELD AT PHILADELPHIA,

FOR PROMOTING USEFUL KNOWLEDGE.

VOL. XVIII.—NEW SERIES.

PUBLISHED BY THE SOCIETY.

Philadelphia:
MacCalla & Company Inc., Printers,
1896.

CONTENTS OF VOL. XVIII.

ARTICLE IV.

OLD BABYLONIAN INSCRIPTIONS CHIEFLY FROM NIPPUR.

PART II.

BY H. V. HILPRECHT, PH.D., D.D.,

Professor of Assyrian and Comparative Semitic Philology and Curator of the Babylonian Museum in the University of Pennsylvania.

Read before the American Philosophical Society, January 17, 1896.

PREFACE.

The publication of the history of the American Expedition to Nuffar, announced in the Preface to the first part of the present work, has been delayed by unforeseen circumstances. In view of the increased interest[1] in these excavations, it seems now necessary to summarize the principal results[2] and submit them to a wider circle of students.

The expedition left America in the summer, 1888, and has continued to the present day, with but short intervals required for the welfare and temporary rest of the members in the field and for replenishing the exhausted stores of the camp. The results obtained have been extraordinary, and, in the opinion of the undersigned editor, have fully repaid the great amount of time and unselfish devotion, the constant sacrifice of health and comfort, and the large pecuniary outlay, which up to date has reached the sum of $70,000. Three periods can be distinguished in the history of the excavations.

[1] Cf. especially the official report on the results of the excavations sent by Hon. A. W. Terrell, the United States Minister in Constantinople, to his government in Washington, summer, 1894.

[2] For details cf. the "Bibliography of the Expedition," in Part I, p. 45. To the list there given may be added Peters, "Some Recent Results of the University of Pennsylvania Excavations at Nippur," in *The American Journal of Archæology* X, pp. 13–46, 352–368 (with copious extracts from Mr. Haynes' weekly reports to the Committee in Philadelphia); Hilprecht, "Aus Briefen an C. Bezold," in *Zeitschrift für Assyriologie* VIII, pp. 386–391 ; *Assyriaca*, Sections I, III–VI. A brief sketch of the history and chief results of the "American Excavations in Nuffar" will be found in Hilprecht, *Recent Research in Bible Lands*, pp. 45–63.

First Campaign, 1888–1889.—Staff: John P. Peters, Director; H. V. Hilprecht and R. F. Harper, Assyriologists; J. H. Haynes, Business Manager, Commissary and Photographer; P. H. Field, Architect; D. Noorian, Interpreter; Bedry Bey, Commissioner of the Ottoman Government.[1] Excavations from February 6 to April 15, 1889, with a maximum force of 200 Arabs. *Principal results:* Trigonometrical survey of the ruins and their surroundings, examination of the whole field by trial trenches, systematic excavations chiefly at III, V, I and X.[2] Many clay coffins examined and photographed. Objects carried away: Over 2000 cuneiform tablets and fragments (among them three dated in the reign of King Ashuretililâni of Assyria), a number of inscribed bricks, terra-cotta brick stamp of Narâm-Sin, fragment of a barrel cylinder of Sargon of Assyria, inscribed stone tablet (Pl. 6), several fragments of inscribed vases (among them two of King Lugalzaggisi of Erech), door-socket of Kurigalzu; c. 25 Hebrew bowls; a large number of stone and terra-cotta vases of various sizes and shapes; terra-cotta images of gods and their ancient moulds; reliefs, figurines and toys in terra-cotta; weapons and utensils in stone and metal; jewelry in gold, silver, copper, bronze and various precious stones; a number of weights, seals and seal cylinders, etc.

Second Campaign, 1889–1890.—Staff: J. P. Peters, Director; J. H. Haynes, Business Manager, Commissary and Photographer; D. Noorian, Interpreter and Superintendent of Workmen; and an Ottoman Commissioner. Excavations from January 14 to May 3, 1890, with a maximum force of 400 Arabs. *Principal results:* Examination of ruins by trial trenches and systematic excavations at III, V and X continued. Row of rooms on the S. E. side of the ziggurrat and shrine of Bur-Sin II excavated. Objects carried away: About 8000 cuneiform tablets and fragments (most of them dated in the reigns of Cassite kings and of rulers of the second dynasty of Ur); a number of new inscribed bricks; 3 brick stamps in terra-cotta and three door-sockets in diorite of Sargon I; 1 brick stamp of Narâm-Sin; 61 inscribed vase fragments of Alusharshid; 2 vase fragments of Entemena of Shirpurla; 1 inscribed unhewn marble block and several vase fragments of Lugalkigubnidudu; a few vase fragments of Lugalzaggisi; 2 door-sockets in diorite of Bur-Sin II; over 100 inscribed votive axes, knobs, intaglios, etc., presented to the temple by Cassite kings; c. 75 Hebrew and other inscribed bowls; 1 enameled clay coffin and many other antiquities similar in character to those excavated during the first campaign but in greater number.

[1] D. G. Prince, of New York, was the eighth member of the expedition, but during the march across the Syrian desert he fell so seriously sick that he had to be left behind at Bagdad, whence he returned to America.

[2] These numbers refer to the corresponding sections of the ruins, as indicated on the plan published in Part I, Pl. XV.

Third Campaign, 1893–1896.—Staff: J. H. Haynes, Director, etc.; and an Ottoman Commissioner; Joseph A. Meyer, Architect and Draughtsman, from June to November, 1894. Excavations from April 11, 1893, to February 15, 1896 (with an interruption of two months, April 4 to June 4, 1894), with an average force of 50–60 Arabs. *Principal results:* Systematic excavations at III, I, II, VI–X, and searching for the original bed and banks of the Shaṭṭ-en-Nil. Examination of the lowest strata of the temple, three sections excavated down to the water level; critical determination of the different layers on the basis of uncovered pavements and platforms; the later additions to the ziggurrat studied, photographed and, whenever necessary, removed; the preserved portions of Ur-Gur's ziggurrat uncovered on all four sides; systematic study of the ancient system of Babylonian drainage; the two most ancient arches of Babylonia discovered; structures built by Narâm-Sin and pre-Sargonic buildings and vases unearthed; c. 400 tombs of various periods and forms excavated and their contents saved. Objects carried away: About 21,000 cuneiform tablets and fragments (among them contracts dated in the reign of Dungi and of Darius II and Artaxerxes Mnemon); many bricks of Sargon I and Narâm-Sin; the first inscribed brick of Dungi in Nippur; 15 brick stamps of Sargon I, 1 of Narâm-Sin; inscribed torso of a statue in diorite ($\frac{2}{3}$ of life size, c. 3000 B.C.) and fragments of other statues of the same period; incised votive tablet of Ur-Enlil; 3 unfinished marble blocks of Lugal-kigub-nidudu and over 500 vase fragments of pre-Sargonic kings and patesis; c. 60 inscribed vase fragments of Alusharshid, 1 of Sargon, 3 of Entemena; 1 door-socket and 1 votive tablet of Ur-Gur; 1 votive tablet of Dungi; a number of inscribed lapis lazuli discs of Cassite kings; fragment of a barrel cylinder of the Assyrian period; fragments of an Old Babylonian terra-cotta fountain in high relief; water cocks, drain tiles, a collection of representative bricks from all the buildings found in Nippur; c. 50 clay coffins and burial urns, and many other antiquities of a character similar to those excavated during the first two campaigns but in greater number and variety.

With regard to the wealth of its results this Philadelphia expedition takes equal rank with the best sent out from England or France. The systematic and careful manner of laying bare the vast ruins of the temple of Bêl and other buildings in Nuffar, with a view to a complete and connected conception of the whole, is equal to that of Layard and Victor Place in Assyria and something without parallel in previous expeditions to Babylonia. Only an exhaustive study and a systematic publication of selected cuneiform texts, which will finally embrace twelve volumes of two to three parts each, can disclose the manifold character of these documents—syllabaries, letters, chronological lists, historical fragments, astronomical and religious texts, building inscriptions, votive tablets, inventories, tax lists, plans of estates, contracts, etc. The results so far obtained have already proved their great importance in connection with ancient chronology, and the fact that nearly all the periods of Babylonian history are represented by inscriptions from the same ruins will enable us, in these publications, to establish a sure foundation for palæographic research.

Each of the three expeditions which make up this gigantic scientific undertaking has contributed its own peculiar share to the total results obtained. The work of the first, while yielding many inscribed documents, was principally tentative and gave us a clear conception of the grandeur of the work to be done. The second continued in the line of research mapped out by the first, deepened the trenches and gathered a richer harvest in tablets and other inscribed monuments. But the crowning success was reserved for the unselfish devotion and untiring efforts of Haynes, the ideal Babylonian explorer. Before he accomplished his memorable task, even such men as were entitled to an independent opinion, and who themselves had exhibited unusual courage and energy, had regarded it as practically impossible to excavate continuously in the lower regions of Mesopotamia. On the very same ruins of Nippur, situated in the neighborhood of extensive malarial marshes and "amongst the most wild and ignorant Arabs that can be found in this part of Asia,"[1] where Layard himself nearly sacrificed his life in excavating several weeks without success,[2] Haynes has spent almost three years continuously, isolated from all civilized men and most of the time without the comfort of a single companion. It was, indeed, no easy task for any European or American to dwell thirty-four months near these insect-breeding and pestiferous Affej swamps, where the temperature in perfect shade rises to the enormous height of 120° Fahrenheit (= c. 39° Réaumur), where the stifling sand-storms from the desert rob the tent of its shadow and parch the human skin with the heat of a furnace, while the ever-present insects bite and sting and buzz through day and night, while cholera is lurking at the threshold of the camp and treacherous Arabs are planning robbery and murder—and yet during all these wearisome hours to fulfill the duties of three ordinary men. Truly a splendid victory, achieved at innumerable sacrifices and under a burden of labors enough for a giant, in the full significance of the word, a *monumentum cere perennius*.

But I cannot refer to the work and success of the Babylonian Exploration Fund in Philadelphia without saying in sorrow a word of him who laid down his life in the cause of this expedition. Mr. Joseph A. Meyer, a graduate student of the Department of Architecture in the Massachusetts Institute of Technology, in Boston,

---

[1] Layard, *Nineveh and Babylon*, p. 565.

[2] Layard, *l. c.*, pp. 556–562. "On the whole, I am much inclined to question whether extensive excavations carried on at Niffer would produce any very important or interesting results" (p. 562).

had traveled through India, Turkey and other Eastern countries to study the history of architecture to the best advantage. In May, 1894, he met Mr. Haynes in Bagdad and was soon full of enthusiasm and ready to accompany him to the ruins of Nuffar. By his excellent drawings of trenches, buildings and objects he has rendered most valuable service to this expedition. But in December of the same year his weakened frame fell a victim to the autumnal fevers on the border of the marshes, where even before this the Syrian physician of the second campaign and the present writer had absorbed the germs of malignant typhus. In the European cemetery of Bagdad, on the banks of the Tigris, he rests, having fallen a staunch fighter in the cause of science. Even if the sand-storms of the Babylonian plains should efface his solitary grave, what matters it? His bones rest in classic soil, where the cradle of the race once stood, and the history of Assyriology will not omit his name from its pages.

The Old Babylonian cuneiform texts submitted in the following pages have again been copied and prepared by my own hand, in accordance with the principle set forth in the Preface to Part I. The favorable reception which was accorded to the latter by all specialists of Europe and America has convinced me that the method adopted is the correct one. I take this opportunity to express my great regret that this second part of the first volume could not appear at the early date expected. The fact that two consecutive summers and falls were spent in Constantinople, completing the reorganization of the Babylonian Section of the Imperial Museum entrusted to me; that during the same period three more volumes were in the course of preparation, of which one is in print now;[1] that a large portion of the time left by my duties as professor and curator was to be devoted to the interest of the work in the field; that the first two inscriptions published on Pls. 36–42 required more than ordinary time and labor for their restoration from c. 125 exceedingly small fragments; and that, finally, for nearly four months I was deprived of the use of my overtaxed eyes, will, I trust, in some degree explain the reasons for this unavoidable delay. In connection with this statement I regard it my pleasant duty to express my sincere gratitude to George Friebis, M.D., my valued confrère in the American Philosophical Society, for his unceasing interest in the preparation of this volume, manifested by the great amount of time and care he devoted to the restoration of my eyesight.

The publication of this second part, like that of the first, was made possible by the liberality and support of the American Philosophical Society, in whose TRANSACTIONS it appears. To this venerable body as a whole, and to the members of its Publication Committee, and to Secretary Dr. George H. Horn, who facilitated the print-

---

[1] Vol. IX, *Tablets Dated in the Reigns of Darius II and Artaxerxes Mnemon*, prepared in connection with my pupil, Rev. Dr. A. T. Clay, now instructor of Old Testament Theology in Chicago.

ing of this work in the most cordial manner, I return my heartiest thanks and my warm appreciation.

No endeavor has been made to arrange Nos. 86–117 chronologically. Although on palæographic evidence certain periods will be readily recognized in these texts, the cuneiform material of the oldest phase of Babylonian history is still too scanty to allow of a safe and definite discrimination. In order to present the monumental texts from Nippur as completely as possible, the fragment of a large boundary stone now in Berlin has found a place in these pages. For permitting its reproduction and for providing me with an excellent cast of the original, Prof. A. Erman, Director of the Royal Museums, has my warmest thanks. I acknowledge likewise my obligations to Dr. Talcott Williams of Philadelphia and to Rev. Dr. W. Hayes Ward of New York for placing the fragment of a barrel cylinder of Marduk-shâbik-zêrim and the impression of a Babylonian seal cylinder respectively at my disposal. If the text of the latter had been published before, Prof. Sayce would not have drawn his otherwise very natural inference (*The Academy*, Sept. 7, 1895, p. 189) that the Hyksos god Sutekh belongs to the language and people of the Cassites.[1] I do not need to offer an apology for including the large fragment of Narâm-Sin's inscription (No. 120), the only cuneiform tablet found in Palestine (No. 147) and the first document of the time of Marduk-ahê-irba,[2] a member of the Pashe dynasty, in the present series. In view of the great importance which attaches to these monuments, a critical and trustworthy edition of their inscriptions had become a real necessity.

The little legend, No. 131, the translation of which is given in the "Table of Contents," will prove of exceptional value to metrologists. At the same time I call the attention of Assyriologists to the interesting text published on Pl. 63, which was restored from six fragments found among the contents of as many different boxes of tablets.

Nos. 124 and 126, which were copied during the time of the great earthquakes in Constantinople, 1894, belong to the collection designated by me as Coll. Rifat Bey. Together with several hundred other tablets they were presented to the Imperial Ottoman Museum by Rifat Bey, military physician of a garrison stationed in the neigh-

---

[1] Prof. Sayce's view rests on Mr. Pinches's hasty transliteration made in connection with a brief visit to America in 1893 and published in Dr. Ward's *Seal Cylinders and Other Oriental Seals* (Handbook No. 12 of the Metropolitan Museum of Art in New York), No. 391, where the Cassite god *Shugab* (= Nergal, cf. Delitzsch, *Kossäer*, p. 25, l. 12) was transliterated incorrectly by *Shu-tah*. I called Dr. Ward's attention to this apparent mistake and gave the correct reading in my *Assyriaca*, p. 93, note.

[2] A boundary stone. The inscription has suffered much from its long exposure to the rain and sun of Babylonia. The original, which the proprietor kindly permitted me to publish, is in Constantinople. The stone is so important that it should be purchased by an American or European museum. My complete transliteration and translation of this text and of Nos. 151 and 152 will appear in one of the next numbers of *Zeitschrift für Assyriologie*.

borhood of Tello, and were catalogued by the undersigned writer. His Excellency, Dr. Hamdy, Director General, and his accomplished brother, Dr. Halil, Director of the Archæological Museum on the Bosphorus, who in many ways have efficiently promoted the work of the American Expedition, and who by their energetic and inte.ligent efforts have placed the rapidly growing Ottoman Museum on a new, scientific basis, deserve my heartiest thanks for permitting the publication of these texts, and for many other courtesies and personal services rendered during my repeated visits to the East.

For determining the mineralogical character of the several stones, I am greatly indebted to my colleagues, Profs. Drs. E. Smith and A. P. Brown, of the University of Pennsylvania.

The systematic excavations of the last decenniums have revolutionized the study of ancient history and philology, and they have opened to us long-forgotten centuries and millenniums of an eventful past. Hieroglyphics and cuneiform inscriptions were deciphered by human ingenuity, and finally the brilliant reasoning and stupendous assiduity of Jensen in Marburg have forced the "Hittite" sphinx to surrender her long-guarded secret. He who has taken the pains to read and read again and analyze the results of Jensen's extraordinary work critically and *sine ira et studio*, must necessarily arrive at the conclusion as to the general correctness of his system. I am neither a prophet nor the son of a prophet, but I see the day not very far, when the world will wonder—just as we wonder now when we glance back upon the sterile years following Grotefend's great achievement—that at the close of the nineteenth century years could elapse before Jensen's discovery and well-founded structure created any deep interest and received that general attention which it deserves. The beautiful marble slab recently found near Malatia[1] has offered a welcome opportunity to test the validity of his theory. But the great *desideratum* seems to be more material than is at present at our disposal. Excavations in the mounds of Malatia would doubtless yield it. But what European government, what private citizens, will furnish the necessary funds? May the noble example given by a few liberal gentlemen of Philadelphia find a loud echo in other parts of the world, and may the work which they themselves have begun and carried on successfully and systematically for several years in Nippur, never lack that hearty support and enthusiasm which characterized its past history. The high-towering temple of Bêl is worthy of all the time and labor

[1] May 23, 1894, together with two other smaller fragments, and now safely deposited in the Imperial Ottoman Museum. With Hamdy Bey's permission published in Hilprecht, *Recent Research in Bible Lands*, p. 160. Cf. also Hogarth in *Recueil*, XVII, p. 25 f. The inscription cannot be older than 750-700 B.C. The artist took as his motive a hunting scene from the royal palaces of Nineveh. A critical analysis of the well-preserved text will be given by Jensen in the next number of *Recueil*.

and money spent in its excavation.   Though now in ruins, the vast walls of this most ancient sanctuary of Shumer and Akkad still testify to the lofty aspirations of a bygone race, and even in their dreary desolation they seem to reëcho the ancient hymn once chanted in their shadow :

Shadû rabû ^{ilu}*Bêl Imḫarsag*	O great mountain of Bêl, Imkharsag,
sha rêshâshu shamâmi shannâ	whose summit rivals the heavens,
apsû ellim shurshudû ushshûshu	whose foundations are laid in the bright abysmal sea,
ina mâtâti kîma rîmi ekdu rabṣu	resting in the lands as a mighty steer,
ḳarnâshu kîma sharûr ^{ilu}*Shamash shittananbiṭû*	whose horns are gleaming like the radiant sun,
kîma kakkab shamê nabû malû siḫâti.	as the stars of heaven are filled with lustre.

(IV *R*. 27, No. 2, 15-24.)

FEBRUARY 15, 1896.

H. V. HILPRECHT.

INTRODUCTION.

I.

THE LOWEST STRATA OF EKUR.

The vast ruins of the temple of Bêl are situated on the E. side of the now empty bed of the Shatt-en-Nîl, which divided the ancient city of Nippur into two distinct parts.[1] At various times the space occupied by each of the two quarters differed in size considerably from the other. Only during the last centuries before the Christian era, when the temple for the last time had been restored and enlarged on a truly grand scale by a king whose name is still shrouded in mystery,[2] both sides had nearly the same extent. This became evident from an examination of the trial trenches cut in different parts of the present ruins and from a study of the literary documents and other antiquities obtained from their various strata. As long, however, as the temple of Bêl existed, the E. quarter of the city played the more important *rôle* in the history of Nippur.

Out of the midst of collapsed walls and buried houses, which originally encompassed the sanctuary of Bêl on all four sides and formed an integral part of the large temple enclosure, there rises a conical mound to the height of 29 m.[3] above the plain and 15 m. above the mass of the surrounding *débris*. It is called to-day *Bint-el-Amîr* ("daughter of the prince")[4] by the Arabs of the neighborhood and covers the ruins of the ancient *ziggurratu* or stage tower of Nippur, named *Imĝarsag*[5] or *Sagash*[6] in the cuneiform

---

[1] Layard (*Nineveh and Babylon*, p. 551) and Loftus (*Travels and Researches*, p. 101) stated this fact clearly. Notwithstanding their accurate description, on most of our modern maps the site of the city is given inaccurately by being confined to the E. side of the canal.

[2] He cannot have lived earlier than c. 500 B.C., and probably later.

[3] Loftus's estimate of seventy feet (*l. c.*, p. 101) is too low.

[4] Layard, *l. c.*, p. 557. Cf. Loftus, *l. c.*, pp. 102f.

[5] "Mountain of heaven," pronounced later *Imursag*. Cf. Jensen in Schrader's *Keilinschriftliche Bibliothek* III, Part 1, p. 22, note 5, and Hommel, *Sumerische Lesestücke*, p. 26, No. 306.

[6] "High towering" (on the ending *sh* cf. Hommel, *l. c.*, p. 141, 2a). Cf. II R. 50, 5–6 a, b. A third name existed but is broken away on this tablet (4 a). For *Imĝarsag* cf. also IV R. 27, No. 2, 15 and 17.

inscriptions (cf. Pls. XXIX and XXX). A number of Babylonian kings[1] applied themselves to the care of this temple by building new shrines, restoring old walls and repairing the numerous drains and pavements of the large complex, known under the name of *Ekur* ("mountain house").[2] But the three great monarchs who within the last three millenniums before Christ, above all others,[3] devoted their time and energy to a systematic restoration and enlargement of the ziggurrat and its surroundings, and who accordingly have left considerable traces of their activity in Nuffar,[4] are Ashurbânapal (668–626 B.C.),[5] Kadashman-Turgu (c. 1250 B.C.)[6] and Ur-Gur (c. 2800 B.C.).[7] The structures of each of these builders have been, one after the other, cleared, measured, photographed and examined in all their details by Mr. Haynes, the intrepid and successful director of the American expedition during the last four years. He is soon expected to communicate the complete results of his work, illustrated by numerous drawings and engravings, in Series B of the present publication. Therefore, referring all Assyriologists to this proposed exhaustive treatise on the history of the excavations, I confine myself to a brief examination of the lowest strata of ancient Ekur, which will enable us to gain a clearer conception of the earliest phase of Babylonian history. Whenever it seems essential, Haynes's own words will be quoted from his excellent weekly reports to the Committee in Philadelphia.

UR-GUR.

At the time of King Ur-Gur the ziggurrat of Nippur stood on the N.-W. edge of an immense platform, which formed the pavement of the entire temple enclosure. It was laid about 2.5 m. above the present level of the plain and had an average thickness of 2.40 m. In size,[8] color and texture the sun-dried and uninscribed bricks of

---

[1] Among them **Dungi** (Pl. 52, No. 123, cf. his brick legend in Part III of the present work), **Ur-Ninib** (Pl. 18, No. 10, and Pl. XXIII, No. 65), **Bur-Sin I** (Pl. 11, No. 19), **Ishme-Dagân** (Pl. 9, No. 17, cf. his brick legend in Part III), **Bur-Sin II** (Pls. 12f., Nos. 20-22), **Kurigalzu** (Pl. 20, No. 38), **Ramm^ n-shumuṣur** (Pl. 28, No. 81), **Esarhaddon** (cf. Vol. X of the present work and Hilprecht in Z. A., VIII, pp. 390f.). As to the earliest builders cf. below.

[2] Cf. Pl. 1, No. 1, 8 ; Pl. 2, No. 2, 10 ; Pl. 20, No. 38, 7 ; Pl. 28, No. 81, 8 ; Pl. 29, No. 82, 8 ; Pl. 51, No. 121, 8 ; also Jensen, *Kosmologie*, pp. 185ff.

[3] With the exception of the unknown builder above referred to, who enlarged the base of the early ziggurrat considerably and changed its form entirely by adding a peculiar cruciform structure (each arm being 16.48 m. long by 6.16 m. wide) to the centre of its four sides. Each side appeared to have a gigantic wing.

[4] Cf. Part I, p. 5, note, and Nöldeke in Hilprecht, *Assyriaca*, p. 86, note 1.

[5] Cf. Pl. 29, No. 82, and Hilprecht in Z. A., VIII, pp. 389ff.

[6] Cf. Pl. 24, No. 8, 8. His brick legend will be published in Part III.

[7] Cf. I R. 1, No. 8f., and Pls. 51f. of the present work.

[8] 23 × 15 4 × 7.7 cm., practically the same size as Ur-Gur's bricks found in the *Buwariyya* of Warka. Cf. Loftus, l. c., p. 168.

this pavement are identical with the mass of crude bricks forming the body of the ziggurrat, while in size and general appearance they closely resemble the burned bricks which bear the name of Ur-Gur. The natural inference would be that Ur-Gur himself erected this large terrace to serve as a solid foundation for his lofty temple. Yet so long as the inside of the massive ruins has not been thoroughly explored, there remains a slight possibility that the body of the ziggurrat and the pavement existed before Ur-Gur, and that this king only repaired and restored an older building, using in the manufacture of his bricks the mould of his predecessor. On the basis of the present almost convincing evidence, however, I favor the former view and, with Haynes, doubt very much whether before Ur-Gur's time a ziggurrat existed in ancient Nippur.[1]

The base of Ur-Gur's ziggurrat formed a right-angled parallelogram nearly 59 m. long and 39 m. wide.[2] Its two longest sides faced N.-W. and S.-E. respectively,[3] and the four corners pointed approximately to the four cardinal points.[4] Three of the stages have been traced and exposed (cf. Pl. XXX). It is scarcely possible that formerly other stages existed above.[5] The lowest story was c. $6\frac{1}{3}$ m. high, while the second (receding a little over 4 m. from the edge of the former) and the third are so

[1] The ancient name of the temple, *Ekur*, in use even at Sargon's time, proves nothing against this theory. On the basis of Taylor's, Loftus's and his own excavations, Haynes inclines to the view that Ur-Gur was the first builder of ziggurrats in Babylonia. As these two English excavators however did not examine the strata below Ur-Gur's terraces, it will be wiser to suspend our judgment for the present, although the absence of a ziggurrat in Tello favors Haynes's view.

[2] In size practically identical with Ur-Gur's structure in Muqayyar (ratio of 3 : 2). Cf. Loftus, *l. c.*, p. 129.

[3] The longest sides of the ziggurrat in Ur faced N. E. and S. W. respectively. Cf. Loftus, *l. c.*, p. 128.

[4] "The N. corner is 12° E. of N." (Peters in *The American Journal of Archæology*, X, p. 18). The Babylonian orientation was influenced by the course of the Euphrates and Tigris, as the Egyptian by the trend of the Nile valley (Hagen in *Beiträge zur Assyriologie* II, p. 246, note). The Assyrian word for "North," *ish(l)tânu*, means "No. I." From this fact, in connection with the observation that in the Babylonian contract literature, etc., in most cases the **upper** smaller side (or front) of a field faces N., it follows that the Babylonians looked towards N. in determining the four cardinal points, and accordingly could not very well designate "West" by a word which means originally "back side" (Delitzsch, *Assyrisches Handwörterbuch*, p. 44f., and Schrader in *Sitzungsberichte der Königl. Preussisch. Akademie der Wissenschaften*, 1894, p. 1301) like the Hebrews, who faced E. Besides, it is grammatically scarcely correct to derive אורי׳ה, a Babylonian loan-word in the Talmud, from a supposed Babylonian *aḫa(u)rru* instead of *avurru* [for this very reason I read the bird mentioned in II *R.* 37, 13 e. f., not *a-ḫar-shu-nu* (Delitzsch, *l. c.*, p. 45) but *a-**mur**-shu-nu*=אורשנא (cf. Halévy in *Revue Sémitique* III, p. 91)]. Consequently the only possible reading is *am(v)urru*, "West," as proposed by Delattre, in view of *mâtuA-mu ri* and *âluA-mu-ur-ra* in the Tell-el-Amarna tablets (cf. also a Babylonian (sic!) village or town *A-mu-ur-ri-iki* in Meissner, *Beiträge zum Altbabylonischen Privatrecht*, No. 42, 1 and 21). Independently a similar result was reached by Hommel in *Zeitschrift der Deutschen Morgenländischen Gesellschaft* XLIX, p. 524, note 3.

[5] No trace of a fourth story could be discovered, and the accumulation of *débris* on the top of Bint-el-Amîr is not large enough to warrant the assumption of more than three stages. In Ur Loftus discovered but two distinct stages (*l. c.*, p. 128).

utterly ruined that the original dimensions can no more be given.[1] The whole ziggurrat appears like an immense altar, in shape and construction resembling a smaller one
discovered in a building to the S.-W. of the temple.

As stated above, the body (and faces) of the ziggurrat consist of small, crude
bricks,[2] with the exception of the S.-E. side of the lowest stage, which had an external facing of burned bricks of the same size.[3] To preserve such a structure for any
length of time it was necessary to provide it with ample and substantial drainage.
Thanks to the untiring efforts of Haynes, who for the first time examined the ancient
Babylonian system of canalisation critically, we learn that the ziggurrat of Nippur
had water conduits of baked brick[4] in the centre of each of the three unprotected
sides. They were found in the lower stage and possibly existed also in the upper[5]
ruined portions. On all four sides around the base of the walls was a plaster of bitumen,[6] 2.75 cm. wide and gradually sloping outward from the ziggurrat towards a
gutter, which carried the water away (cf. Pl. XXIX, No. 74).[7] By this very simple
arrangement the falling rain was conducted to a safe distance and the unbaked brick
foundations were thoroughly protected.

Unlike the ziggurrat of Sin in Ur, which had its entrance on the N.-E. side,[8] the
ascent to the different stages in Nippur was at the S.-E. Two walls of burned bricks,[9]
3.40 m. high, 16.32 m. long and 7 m. distant from each other, ran nearly parallel,[10] at

---

[1] The surface of these stages "was covered with a very tenacious plaster of clay mixed with cut straw," in order
to protect them against storm and rain. "In places this plaster is still perfect, while in other places several coatings
are visible, plainly showing that from time to time the faces of the ziggurrat were replastered" (Haynes, Report of
Sept. 1, 1894).

[2] Cf. above, p. 230, note 8, "Traces of decayed straw were discovered in these bricks" (Haynes, Report of Feb.
9, 1895).

[3] In Ur the exterior of the whole lower story was faced by Ur-Gur with baked bricks (Loftus, l. c., pp. 129f.),
while in Warka "unlike other Babylonian structures" the lower stage of the Buwariyya "is without any external
acing of kiln-baked brickwork" (Loftus, l. c., p. 167).

[4] Each c. 1 m. wide by 3.25 deep. To judge from the height of the "buttresses" in Warka, the true meaning of
which Loftus failed to recognize, the lowest stage of the Buwariyya had the same height as that of the ziggurrat of
Nippur. Cf. Loftus, l. c., p. 169.

[5] Cf. Loftus, l. c., p. 129.

[6] This plaster rested upon "a level pavement of two courses of bricks also laid in bitumen, and was 28 cm. thick
where it flanked the walls, and 7.7 cm. at its outer edge" (Haynes, Report of Feb. 10, 1894).

[7] The projecting casing wall at the base (1.38 m. high) consists of sixteen courses of (stamped) bricks and was
built by Kadashman-Turgu around the three unprotected sides of the ziggurrat. In the middle distance of the picture
is seen a section of the latest crude brick superstructure (cf. above, p. 230 and note 3) with a tunnel tracing the face of
the lowest stage of Ur-Gur's and Kadashman-Turgu's ziggurrat.

[8] Loftus, l. c., p. 129.

[9] Many of which were stamped with Ur-Gur's well-known legend I R. 1, No. 9.

[10] Where they joined the wall of the ziggurrat the distance between them (7 m.) was 1.65 m. greater than at their
outer end.

right angles from the face of the ziggurrat, into the large open court, which extended to the great fortification of the temple. This causeway[1] was filled up with crude bricks of the same size and mould and formed a kind of elevated platform, from which apparently steps, no longer in existence, led up to the top of the ziggurrat and down into the open court in front of it.

The whole temple enclosure was surrounded by a large inner and outer wall built of sun-dried bricks. To the N.-W. of Ekur "30 courses of these bricks are still plainly visible."[2] They compose the ridge of the outer wall and, like the pavement of Ur-Gur's ziggurrat, rest on an older foundation. The complete excavation of the inner wall will be undertaken in connection with the systematic examination and removal of the ruins around the ziggurrat.

SARGON AND NARÂM-SIN.

Immediately below "the crude brick platform of Ur-Gur," under the E. corner of the ziggurrat, was another pavement consisting of two courses of burned bricks of uniform size and mould.[3] Each brick measures c. 50 cm. in square and is 8 cm. thick. This enormous size is quite unique among the more than twenty-five different forms of bricks used in ancient Nippur, and enables us to determine the approximate date of other structures built of similar material in other parts of the city. Fortunately most bricks of this pavement are stamped. A number of them contain the well-known inscription of **Shargâni-shar-âli,** while the rest bears the briefer legend of **Narâm-Sin** (Part I, Pls. 3 and II). This fact is significant. As both kings used the same peculiar bricks, which were never employed again in the buildings of Nippur, and as they are found near together and intermingled in both courses of the same pavement, the two men must necessarily be closely associated with each other. This ancient brick pavement becomes therefore a new and important link in the chain of my arguments in favor of the identity of Shargâni-shar-âli[4] with Sargon I, father of

---

[1] Both the walls of the causeway and those of the ziggurrat were battered, the batter of the former (1 : 8) being exactly half the batter of the latter (1 : 4), according to Haynes's Report of Feb. 9, 1895. Cf. Loftus, *l. c.,* p. 128.

[2] Haynes, Report of Sept. 8, 1894.

[3] Niebuhr's very recent remarks on the historicity of Sargon I and Narâm-Sin (*Chronologie der Geschichte Israels, Ægyptens, Babyloniens und Assyriens,* Leipzig, 1896, p. 75) should never have been made after the publication of their inscriptions in the first part of the present work. His insinuations against the priests of Nippur read like a carnival joke, in the light of the facts presented in the following sketch.

[4] Oppert's proposed reading of this name as *Bingani sar-iris* (*Revue d'Assyriologie* III, pp. 25f.) is impossible and was declined in *Assyriaca,* p. 30, note 1. The original picture of the sign *Shar* in our name is not "l'hiéroglyphe de l'arbre en feuilles" (Oppert, *l. c.*), but an enclosed piece of land covered with plants, in other words a plantation, garden, orchard (*kirû*). Cf. Bertin, *Origin and Development of the Cuneiform Syllabary,* p. 7.

Narâm-Sin [1] (Part I, pp. 16–19). It was apparently laid by Sargon and relaid by his son, Narâm-Sin, who utilized part of his father's bricks, and it must therefore be recognized as the true level of the Sargon dynasty in the lower strata of the temple at Nuffar. No bricks of either of the two kings have been found below it, nor in fact any other inscribed objects that can be referred to them.[2] But another, even more powerful witness of Narâm-Sin's activity in Nippur [3] has arisen from some ruins in the neighborhood of Ekur.

On the plan of Nuffar published in Part I, Pl. XV, a ridge of low insignificant-looking mounds to the N.-W. of the temple [4] is marked VII. They represent a portion of *Nimit-Marduk*, the outer wall of the city.[5] Its upper part, as stated above, was constructed by Ur-Gur. During the summer of 1895 Mr. Haynes excavated the lower part of this rampart. He selected a piece of 10 m. in length and soon afterwards reported the following surprising results. The foundation of the wall was placed on solid clay c. $\frac{2}{3}$ m. below the water level or c. 5 m. below the plain of the desert. It was "built of worked clay mixed with cut straw and laid up *en masse* with roughly sloping or battered sides" to a total height of c. 5.5 m. Upon the top of this large base, which is c. 13.75 m. wide, a wall of the same enormous width, made of sun-dried

---

[1] More recently (*Altorientalische Forschungen* III, p. 238) Winckler refers to Shargâni-shar-âli as the possible historical basis of "the mythical Sargon of Agade." I trust the day is not very far when he will regard Sargon as historical and identical with Shargâni-shar-âli, as I do.

[2] The brick stamp of Sargon, mentioned below, p. 243, as having been unearthed underneath the wall of Ur-Gur's archive, indicates that this underground archive or cellar existed at Sargon's time at that very spot and was rebuilt by Ur-Gur.

[3] Inscribed burned bricks of Narâm-Sin were also found in mound X, on the W. bank of the Shatt en-Nîl at a very low level. All the stamped bricks of Narâm-Sin "show evident traces of red coloring on their under or inscribed face" (Haynes, Report of Nov. 24, 1894).

[4] Originally these mounds continued a little farther N. W. than they can be traced on the map, until suddenly they turned to the W., reaching the Shatt en-Nîl apparently not far from II. A large open space, "414 m. long by 276 m. wide and covering more than 26 acres of ground," was enclosed by this wall, by the mounds called VIII and by the temple complex (III). As far as the present evidence goes, this court was never occupied by any brick buildings. Its real purpose can therefore only be surmised. According to Haynes (Report of August 3, 1895) it served as a caravanserai for the accommodation and safety of pilgrims and their animals. Such a view is possible, but it seems to me more probable to regard this enclosed place as a court where the numerous cattle, sheep, etc., received by the temple administration as regular income and for special sacrifices, were kept and sheltered. Perhaps it served both purposes. Besides in the time of war the inhabitants of Nippur readily found a safe refuge behind its walls. On the N. E. side of this court, "at the foot of the enclosing wall, a bubbling spring was discovered. On either side of the spring are still seen the brick platforms and curbs where the water pots rested." From the size of the bricks, which "appear to be the half bricks of Narâm-Sin," the spring existed at the time of this great builder. "After the court had become filled to a depth of about 1 m, a diagonal wall of burned bricks, 5½ m. long, six courses high, placed on a raised base of clay, was built before the spring to divert the course of drifting sand and *débris* from the court."

[5] Cf. II R. 50, 29 a, b. The inner fortification (*dûru*) was called *Imgur-Marduk* (*ibidem*, 28 a, b). Cf. Delitzsch, *Wo lag das Paradies?* p. 221. Both names seem to be of comparatively late date and cannot be applied to Narâm-Sin's fortifications. According to II R. 50, 30f, a, b, two other names existed for the outer wall (*shalḫû*).

bricks, was raised to an unknown height.[1] We may well ask in amazement, Who was the builder of this gigantic wall, constructed, as it seems, *ana ûm ṣâte?* Nobody else than the great Narâm-Sin, whom Niebuhr of Berlin finds hard to regard as a historical person! Perhaps this scholar will now release me from presenting "wirkliche Inschriften politischer und als solcher glaubhafter[2] Natur, damit man ihrer [namely, Sargon's and Narâm-Sin's] einstmaligen Existenz vollkommen traue."[3] The bricks had exactly the same abnormal size as the burned bricks of the pavement below the ziggurrat and, in addition, although unbaked, bore Narâm-Sin's usual stamped inscription of three lines. "They are dark gray in color, firm in texture and of regular form. In quality they are unsurpassed by the work of any later king, constituting by far the most solid and tenacious mass of unbaked brick that we have ever attempted to cut our way through."[4] A large number of "solid and hollow terra-cotta cones in great variety of form and color,"[5] and many fragments of water spouts were found in the *débris* at the bottom of the decaying wall. The former, as in Erech,[6] were used for decoration, the latter apparently for the drainage of the rampart.[7] Possibly there were buildings of some kind on the spacious and airy summit of the wall,[8] although nothing points definitely to their previous existence.

[1] I have summarized the details of Haynes's report, according to which the original base was c. 5 m. high and c. 10.75 m. wide. "Directly upon this foundation Narâm-Sin began to build his wall, 10.75 m. wide and six courses high. For some reason unknown to us, the builder changed his plan at this point and widened the wall by an addition of c. 3 m. in thickness to the inner face of the wall, making the entire thickness or width of the wall c. 13.75 m. This addition, like the original foundation, was built of worked clay mixed with cut straw, and from the clay bed was built up to the top of the moulded brick wall, making a new and wider base, c. 5.5 m. high by c. 13.75 m. wide. Upon this new and widened base a new wall of equal width was built by Narâm-Sin, whose stamped bricks attest his workmanship. In the construction of the original base, c. 5 m. high and c. 10.75 m. wide, there is nothing to furnish a clue to its authorship" (Report of August 3, 1895). In the same letter Haynes argues very plausibly, as follows : " Had the superstructure been built upon the original base, as it was begun, it would naturally appear that the entire structure from its foundation was the work of Narâm-Sin ; yet because Narâm-Sin changed the proportions of the wall, it may with some show of reason be assumed that Narâm-Sin himself began to build upon the foundation of a predecessor, perhaps of his father Sargon, with the intention of completing the original design, and that his own ideas then began to fix upon a different or at least upon a larger plan requiring a wider base to build upon."

[2] I am afraid Niebuhr's use of " politisch " und " glaubhaft " as two corresponding terms is very " unhistorisch." Apparently he has a very curious conception of the significance of an inscribed Babylonian brick as a historical document over against the "political inscriptions" too often subjectively colored. Cf. Maspero, *The Dawn of Civilization*, p. 626, with whom I agree.

[3] Carl Niebuhr, *l. c.*, p. 75.

[4] Haynes, Report of Sept. 8, 1895.

[5] "Red and black color are abundant. The hollow cones are of larger size than the solid cones " (Report of July 27, 1895).

[6] Cf. Loftus, *l. c.*, p. 187ff.

[7] It is doubtful whether the cones and spouts belonged to Narâm-Sin's or Ur-Gur's structure ; the water spouts point to the time of the former, however.

[8] Haynes inclines strongly to the view that there existed "a tier of rooms flush with the outer face of the wall, and a broad terrace before them overlooking the great enclosure " (Report of Aug. 3, 1895). This view is closely

The construction of so gigantic a fortification by Narâm-Sin proves the political importance of Nippur at an early time, and reveals, in its own peculiar way, the religious influence which Ekur exercised in the ancient history of the country. A number of scattered references in the oldest cuneiform inscriptions extant—as, *e. g.*, the fact that the supreme god of Lagash is called *gud Inlil* by several kings and governors of Tello,[1] that Edingiranagin[2] bears the title *mupada Inlila-ge*, that Urukagina[3] as well as Entemena[4] built a shrine to *Inlil*, that the rulers of Kish,[5] Erech[6] and of other early Babylonian centres,[7] who lived about the period of the kings of Shirpurla, paid their respect to Bêl, repeatedly making valuable offerings and numerous endowments, and claimed as *patesi gal Inlila*[8] the right of chief officer in his sanctuary and domain— and the interesting passage in the bilingual text of the creation story,[9] where Nippur seems to be regarded as the oldest city of Babylonia, find a welcome confirmation in the results obtained by our systematic excavations.

A comparatively small portion of the enormous temple area has so far been thoroughly examined, although for more than five years the constant hard labor of fifty to four hundred Arabic workmen has been devoted to its exploration. The results have already been extraordinary; they will become more so when our work shall be completed. That no independent buildings of Sargon have as yet been discovered will be partly explained in the light of the statement just made. The large number of Sargon's brick stamps[10] excavated at different times chiefly within the temple enclosure,

connected with his theory as to the use of the court, above referred to. "In a hot country, infested with robbers and swarming with insects, the rooms on the wall and the terrace in front of them would have offered admirable sleeping quarters for the hosts of pilgrims at Bêl's most famous shrine (*ibidem*)."

[1] *E. g*, by **Urukagina** [De Sarzec, *Découvertes en Chaldée*, p. XXX, squeeze (cf. p. 109f.), col. I, 2 ; and Pl. 5, No. 1, 2f. (also Amiaud, on p. XXX)], **Enanatuma I** [inscription published by Heuzey in *Revue d'Assyriologie* III, p. 3', 2], **Entemena** [De Sarzec, *l. c.*, Pl. 31, No. 3, col. I, 2 ; and *Revue d'Assyriologie* II, p. 148, col. I, 2], **Enanatuma II** [De Sarzec, *l. c.*, Pl. 6, No. 4, 2].

[2] De Sarzec, *l. c.*, Pl. 31, No. 2, col. I, 5f. (cf. *Revue d'Assyriologie* II, p. 81).

[3] De Sarzec, *l. c.*, Pl. 5, No. 1, 35–38 ; Pl. 32, col. III, 1–3 ; squeeze (p. XXX), col. III, 7–9.

[4] De Sarzec in *Revue d'Assyriologie* II, p. 149, col. IV, 4–7 (to be supplemented by De Sarzec, *Découvertes*, passages quoted in the preceding note).

[5] Hilprecht, *Old Babylonian Inscriptions*, Part II, Pl. 43, No. 3. Cf. Pl. 46, No. 108.

[6] Hilprecht, *l. c.*, Pls. 38–42, No. 87.

[7] *E. g.*, Ur, cf. Hilprecht, *l. c.*, Pls. 36f., No. 86 ; Pl. 42, No. 88 and No. 89. Cf. also Pl. 42, No. 90 ; Pl. 43, Nos. 91f.

[8] Lugalzaggisi. Cf. Hilprecht, *l. c.*, Pl. 38, No. 87, col. I, 15f.

[9] Pinches in *Records of the Past²*, Vol. VI, p. 109, 6.

[10] Not less than eighteen (either whole or fragmentary) terra-cotta stamps have been unearthed, seven of them within one fortnight in December, 1895. Most of them are without handles. Apparently several broke while in use at Sargon's time and were then thrown away. Others were doubtless broken intentionally in connection with the disastrous event mentioned below, p. 244.

his stamped bricks[1] found under the platform of Ur-Gur, and the regular title *bâni*[2] *Ekur bit Bêl in Nippur* occurring in all his inscriptions from Nuffar[3] indicate that important structures, similar to those of his son, must have existed in some part of these high and extended accumulations. The perplexing question is, at which particular spot have we to search for them? And shall we ever really find them? Just as the bricks of Ur-Gur lie directly upon the splendid structure of Narâm-Sin in the large enclosing wall (*Nimit-Marduk*), so "the great crude brick platform of Ur-Gur's ziggurrat practically rests upon Narâm-Sin's pavement."[4] This fact is of importance, for we draw the natural conclusion from it that all the buildings that once stood upon this latter pavement were razed by Ur-Gur, in order to obtain a level ground for his own extended brick pavement, which served as the new foundation for Ekur.

THE PRE-SARGONIC PERIOD.

The average accumulations of *débris* above the pavement of Narâm-Sin measure a little over 11 m. in height and cover about 4000 years of Babylonian history. Have any traces of an earlier temple beneath the pavement of the Sargon dynasty been found in Nuffar? Several sections on the S.-E. side of the ziggurrat have been excavated by Mr. Haynes down to the water level.[5] I am therefore fully prepared to make the following statement, which will sound almost like a fairy tale in the ears of Assyriologists and historians who have been accustomed to regard the kingdom of Sargon as legendary and the person of Narâm-Sin as the utmost limit of our knowledge of ancient Babylonian history. The accumulations of *débris* from ruined buildings, partly preserved drains, broken pottery and many other remnants of human civilization between Narâm-Sin's platform and the virgin soil below, are not less than 9.25 m. The age of these ruins and what they contain can only be conjectured at the present

---

[1] The fragment of the first Sargon brick excavated in Nuffar at the beginning of 1894 is published on Pl. XXI, No. 63. It proves that Sargon did not only stamp his legend upon the bricks but sometimes wrote it. For a stamped specimen cf. Part III.

[2] Written *ba-GIM* = (*ba-*)*bâni* or (*ba-*)*bân*, in other words expressed by an ideogram and preceding phonetic complement (the earliest example of this kind in Semitic cuneiform texts). Cf. Hilprecht, *Assyriaca*, p. 70, note (end). Examples for this peculiar use of a phonetic complement are extremely rare and will be found in *Assyriaca*, Part II.

[3] Pls. 1-3, Nos. 1-3.

[4] Haynes, Report of Aug. 3, 1895. In advance I warn all those who seem to know Babylonian chronology better (?!) than King Nabonidos of Babylon, not to use this fact against the king's 3200 years, and to keep in mind that also Ur-Gur, Kadashman-Turgu and Ashurbânapal follow each other immediately in their work at the ziggurrat.

[5] To illustrate the amount of time, patience and labor needed for the systematic exploration of these lowest strata, it may be mentioned that one of the sections excavated contained "more than 60,000 cubic feet" of earth, which had to be carried away in basketfuls a distance of 120 m. and at the same time to be raised to a height of 15-24 m. Haynes, Report of Oct. 5, 1895.

time. But as no evidence of an ancient ziggurrat previous to Ur-Gur and Narâm-Sin has been discovered, the accumulations must have necessarily been slower and presuppose a longer period than elapsed between Narâm-Sin and the final destruction of Ekur in the first post-Christian millennium. I do not hesitate, therefore, to date the founding of the temple of Bêl and the first settlements in Nippur somewhere between 6000 and 7000 B.C.,[1] possibly even earlier. I cannot do better than repeat Haynes' own words, written out of the depth of this most ancient sanctuary of the world so far known : "We must cease to apply the adjective earliest to the time of Sargon or to any age or epoch within 1000 years of his advanced civilization."[2] "The golden age of Babylonian history seems to include the reign of Sargon and of Ur-Gur."[3]

Somewhat below the pavement of Narâm-Sin, between the entrance to the ziggurrat and the E. corner, stood an altar of sun-dried brick, facing S.-E. and 4 m. long by 2.46 m. wide. The upper surface of this altar[4] was surrounded by a rim of bitumen (18 cm. high), and was covered with a layer of white ashes (6.5 cm. thick), doubtless the remnant of burned sacrifices. To the S.-W. of it Haynes discovered a kind of bin built of crude brick and likewise filled with (black and white) ashes to the depth of c. 30 cm.[5] At a distance of nearly 2 m. from the altar (in front of it) and c. 1.25 m. below the top was a low wall of bricks, whose limits have not yet been found. Apparently it marked a sacred enclosure around the altar, for it extended far under the pavement of Narâm-Sin[6] and reappeared under the W. corner of the ziggurrat.[7] The bricks of which this curb was built are plano-convex in form.[8] They are laid in mud seven courses (= 45 cm.) high,[9] the convex surface, which is "curiously creased lengthwise," being placed upward in the wall.

At a distance of 4.62 m. outside of this low enclosure and c. 36 cm. below its bottom stood a large open vase in terra-cotta with rope pattern[10] (cf. Pl. XXVII, No. 72). It will serve as an excellent specimen of early Babylonian pottery in the fifth millennium before Christ. Undisturbed by the hands of later builders, it had remained

---

[1] A similar conclusion was reached by Peters in *The American Journal of Archæology* X, pp. 45f.

[2] Report of August 30, 1895.

[3] Report of August 3, 1895.

[4] Which was 0.92 m. below the level of Narâm-Sin's pavement.

[5] Haynes, Report of Feb. 17, 1894 (also Aug. 24, 1895). Haynes's chemical analysis of the white ashes showed evident traces of bones.

[6] The facts concerning this curb have been gathered from Haynes's Reports of Feb. 17 and March 17, 1894; Aug. 3, 1895.

[7] Cf. Peters, *The American Journal of Archæology* X, pp. 31 and 44.

[8] With an average length and breadth of 24.5 × 18 cm.

[9] "Being placed lengthwise and crosswise in alternate courses" (Haynes, Report of March 17, 1894).

[10] Haynes, Report of Aug. 24, 1895.

in its original upright position for more than 6000 years, and it was buried under a mass of earth and *débris* long before Sargon I was born and Narâm-Sin fortified the temple of Nippur.[1]

A second vase of similar size but different pattern[2] was discovered 77 cm. below the former and nearly double the distance from the ancient brick curb. There is little doubt in my mind that both vases, which stood in front of the altar, on its S.-S.-E. side, one behind the other as one approached it, served some common purpose in connection with the temple service at the pre-Sargonic time.

Another section of earth adjoining the excavation which had yielded these remarkable results was removed by Haynes.

To the S.-E. of the altar described above, almost exactly under the E. corner of Ur-Gur's ziggurrat and immediately below the pavement of Narâm-Sin, stood another interesting structure.[3] It is 3.38 m. high,[4] 7 m. square, "with a symmetrical and double reëntrant angle at its northern corner and built up solidly like a tower." Its splendid walls, which exhibit no trace of a door or opening of any kind, are made of large unbaked bricks of tenacious clay[5] somewhat smaller in size than those of Narâm-Sin's rampart. While examining the surroundings of this building, Haynes found ten basketfuls of archaic water vents and fragments thereof on its S.-E. side and on a level with its foundation. His curiosity was aroused at once, and after a brief search underneath the spot where the greatest number of these terra-cotta vents and cocks had been gathered, he came upon a drain which extended obliquely under the entire breadth of this edifice. At its outer or discharging orifice he found the most ancient keystone arch yet known in the history of architecture. The question once asked by Perrot and Chipiez[6] and answered by them with a "probably not," has been definitely decided by the American expedition in favor of ancient Chaldæa. The bottom of this valuable witness of pre-Sargonic civilization[7] was c. 7 m. below the level of Ur-Gur's crude brick platform, 4.57 m. below the pavement of Narâm-Sin, and 1.25 m. below the foundations of the aforesaid building. The arch is 71 cm. high, elliptical in form, and has a span of 51 cm. and a rise of 38 cm. Cf. Pl. XXVIII,

[1] It stood 3.05 m. below the pavement of Narâm-Sin.

[2] In the form of a large jar, its diameter in the centre being larger than that at the top (Haynes, Report of Aug. 24, 1895).

[3] The following facts have been gathered from Haynes's Reports of Oct. 13, Nov. 24, 1894.

[4] Its foundations are therefore 3.38 m. below the level of Narâm-Sin's pavement.

[5] "Thoroughly mixed with finely cut straw and well kneaded."

[6] *A History of Art in Chaldæa and Assyria*, Vol. II, p. 234.

[7] Haynes, Reports of Oct. 13, 20, Nov. 24, 1894; Jan. 12, March 2, 1895.

No. 73.[1] The bricks of which it is constructed are well baked, plano-convex in shape, and laid in clay mortar, the convex side being turned upward. A few months after its discovery the arch was forced out of shape, " probably from the unequal pressure of the settling mass above it, which had been drenched with rain water."

Whether the altar, the two large vases and the massive building, under which the ancient arch was found, had any original connection with each other, is at present impossible to prove. According to my calculations and our latest news from the field of excavation, the bottom of the lower vase and the foundation of the massive building were not on the same level. The difference between them is nearly 0.5 m. As the highest vase, however, stood 77 cm. above the other, and as the section S.-E. from them has not yet been excavated, it is highly probable that a third vase stood at some distance below the second. However this may be, so much we can infer from the facts obtained even now, that an inclined passage from the plain led alongside the two vases to the elevated enclosure around the solitary altar. I am therefore disposed to assign to the tower-like building, the character of which is still shrouded in mystery, the same age as the altar, curb and vases. The keystone arch and drain, on the other hand, are doubtless of a higher antiquity. Whether the 3200 years given by Nabonidos as the period which elapsed between his own government and that of Sargon I, be correct or not, the arch cannot be placed lower than 4000 B.C., and in all probability it is a good deal older.

The two sections which contained all the buildings and objects described above were carried down to the virgin soil, where water stopped our progress. A third section removed in their neighborhood yielded similar results. But it is impossible to enumerate in detail all the antiquities which were uncovered below the S.-E. side of the ziggurrat. The lowest strata did not furnish any treasures similar to those found in the upper layers; they showed a large proportion of black ashes and fine charcoal mingled with earth, but they also produced many smaller objects of great interest and value, especially fragments of copper, bronze and terra-cotta vessels. Several pieces of baked clay steles, bearing human figures in relief upon their surface, will be treated at another place and time.[2] An abundance of fragments of red and black lacquered

[1] A kind of pointed arch of unbaked brick (60 cm. high and 48 cm. wide at the bottom) was found by Haynes in mound X (cf. Pl. XV), on the S. W. side of the canal bed. From the depth in which it was discovered, Haynes reasoned correctly that it was older than 2000 B.C. From the inscribed objects excavated in connection with it, I determined that it must have existed at the time of the dynasty of Isin (c. 2500 B.C.). In all probability it dates back to Ur-Gur's period. For the wall in which this arch is placed was built of the same sun-dried bricks which compose the body of the ziggurrat (Haynes, Reports of April 27, Dec. 21, 1895). For the general form of this pointed arch cf. Perrot and Chipiez, l. c., p. 229, Fig. 92.

[2] One of them was found at a depth of 7 m. below the pavement of Narâm-Sin and 2.44 m. lower than the bottom of the arch, within about 2 m. of the lowest trace of civilization (Haynes, Report of Sept. 7, 1895). Another was discovered 7.70 m. below Narâm-Sin's pavement (Report of Sept. 14, 1895).

pottery was discovered at a depth of 4.6 m. to 8 m. below the pavement of Narâm-Sin.[1] "Had these pieces been found in the higher strata, one would unhesitatingly declare them of Greek origin, or at least ascribe them to the influence of Greek art." For they are, as a rule, of great excellence and in quality far superior to those found in the strata subsequent to the period of Ur-Gur.

The results of our excavations in the deepest strata of Ekur will change the current theory on the origin and antiquity of the arch, will clear our views on the development of pottery in Babylonia, and will throw some welcome rays on one of the darkest periods of history in the valley of the Tigris and Euphrates. But first of all, they again have brought vividly and impressively before our eyes the one fact that Babylonian civilization did not spring into existence as a *deus ex machina ;* that behind Sargon I and Narâm-Sin there lies a long and uninterrupted chain of development covering thousands of years ; and that these two powerful rulers of the fourth millennium before Christ, far from leading us back to " the dawn of civilization," are at the best but two prominent figures from a middle chapter of the early history of Babylonia.

[1] A vase of ordinary gray pottery, 23 cm. high, was found 7.40 m. below this pavement "directly beneath the line of the very ancient curb, and near to a perpendicular let fall from the E. corner of the altar." The stratum which produced this vase, according to Haynes, "was literally filled with potsherds of small size and generally brick red in color" (Report of Sept. 14, 1895).

II.

THE INSCRIBED MONUMENTS OF SARGON'S PREDECESSORS.

Although more than 500[1] mostly fragmentary antiquities of Sargon and his predecessors have been excavated in Nuffar, it may at first seem strange that nearly all of them were discovered out of place, above the platform of Ur-Gur. But if we examine the details more closely, we will easily find the explanation of this remarkable fact. Almost all these monuments that, on the basis of strong palæographic evidence and for various other reasons, must be ascribed to this early phase of Babylonian history,[2] were found in a stratum on the S.-E. side of the ziggurrat, between the facing of the latter and the great fortified wall which surrounded the temple. This stratum varies in thickness. "In some places it lies directly upon the crude brick pavement of Ur-Gur, while in other places it reaches a height of c. 1 m. above this platform."[3] Few of the objects found were whole, the mass of them was broken and evidently broken and scattered around on purpose. Most of the fragments are so small that during the last three years it needed my whole energy and patience, combined with much sacrifice of the eyesight, to restore the important inscriptions published on the following pages (particularly Pls. 36–42). The apparent relation in which this stratum stands to a peculiar building in its immediate neighborhood will furnish the key to the problem.

AN ANCIENT TEMPLE ARCHIVE.

Directly below the great fortification wall of the temple to the S.-E. of the ziggurrat, Mr. Haynes discovered recently a room 11 m. long, 3.54 m. wide and 2.60 m. high. It showed nowhere a door or entrance in its unbroken walls, and there can be no doubt "that the room was a vault entered by means of a ladder, stairway or other perishable passage from above." This structure "was erected on the level of Narâm-Sin's pavement," and yet it was made of the same bricks which compose the

---

[1] Stamped bricks being excluded.

[2] Cf. proof below.

[3] Haynes, Report of Dec. 14, 1895.

body of Ur-Gur's ziggurrat and platform. How is this discrepancy to be explained? By the simple assertion, suggested already by the absence of a door in the walls of the building, that the room was underground, a cellar reaching from the top of Ur-Gur's platform down to the level of Narâm-Sin's pavement.[1] The access from above being on the Ur-Gur level, it is clear that the vault was built by this king himself. Our interest in the unearthed building is still increased by the discovery of another smaller[2] room of exactly the same construction and material below it. Separated from the later vault by a layer of earth and *débris* 60 cm. deep, it lies wholly below the level of Narâm-Sin's platform. In its present form this lower cellar cannot, however, antedate Sargon, nor was it built by this king himself or by his immediate successor. From the fact that the bricks of both rooms are identical "in size, form and general appearance,"[3] and that a brick stamp of Sargon was discovered beneath the foundations of the lower walls, we draw the following conclusions: (1) At the time of Sargon a cellar existed at this very spot, as indicated by the presence of his stamp below the level of his dynasty;[4] (2) Ur-Gur found and used this cellar, but rebuilt it entirely with his own bricks. And as he raised the foundation of his ziggurrat far above the old level, he also raised the walls of the old chamber to the height of his new platform. (3) For some unknown reason—probably because the pressure of the neighboring temple fortifications from above, together with the yearly rains, the principal enemies of Babylonian sun-dried brick structures, had ruined the vault[5]—he changed its foundation afterwards and laid it on a higher level, at the same time widening the space between its two longer walls.

It can be easily proved that this underground building was the ancient storeroom or archive of the temple. "A ledge c. 0.5 m. wide and 0.75 m. above the floor extended entirely around the room, serving as a shelf for the storage of objects in due form and order."[6] "A circular clay tablet together with two small tablets of the ordinary form and five fragments were found on it,"[7] and five brick stamps without handles were lying within its walls. And finally a similar room filled with about 30,000 clay tablets, inscribed pebbles, cylinders, statues, etc., was discovered by de Sarzec, 1894, in a

---

[1] The height of its walls agrees with the distance between the tops of Ur-Gur's and Narâm-Sin's platforms.

[2] It is only 2.15 m. wide, and the walls are 92 cm. high in their present ruined condition.

[3] Haynes, Report of Dec. 14, 1895.

[4] Cf. above, p. 235, note 2.

[5] On this theory it can be easily explained why a few tablets were found on the ledge of the lower room and brick stamps without handles were discovered on the floor of the same room.

[6] Haynes, Report of Dec. 14, 1895. This ledge existed in both chambers. It was built up with the walls and consisted of crude bricks capped by a layer of burned bricks (Report of Dec. 21, 1895).

[7] In the lower vault (Haynes, Report of Dec. 21, 1895). In the midst of this lower chamber was "a hemispherical basin of pottery set in a rim of stone," the original use of which is still unknown (Report of Dec. 14, 1895).

small mound at Tello,[1] by which the true character of our building is determined beyond question. The French explorer was more fortunate than Mr. Haynes in finding his archive undisturbed, but it will always remain a serious loss to science that the contents of the archive of Tello could not have been saved and kept together.[2]

The vault of Nippur had been robbed by barbarians of the third millennium before Christ, as I infer from the following facts and indications:

1. Nearly all the objects above referred to were excavated from a well-defined stratum in the neighborhood of this storeroom. From the position in which they were found, from the fact that none, except door-sockets in diorite, were whole, and from the extraordinarily small size of most fragments, it becomes evident that the contents of the archive were broken and scattered intentionally, as previously stated.

2. Three of the rulers of the dynasty of Isin built at the temple of Nippur,[3] and an inscribed brick of Ur-Ninib was found among the fragments recovered from this stratum. It is therefore clear that the destruction of the vases, brick stamps, etc., did not antedate Ur-Ninib's government. As no document later than his time has been rescued from this stratum, it is also manifest that the deplorable disaster occurred not too long after the overthrow of his dynasty.

3. The archive existed however as late as the second dynasty of Ur. For Bur-Sin II wrote his name on an unhewn block of diorite, presented to Bêl many centuries before by Lugal-kigub-nidudu, a pre-Sargonic[4] king of Ur and Erech, and turned it into a door-socket for his own shrine in Nippur.[5] That the archive could not have been destroyed in the brief interval between Ur-Ninib and Bur-Sin II, so that the latter might have rescued his block from the ruins, results from a study of the general history of that period, however scanty our sources, and of the history of the city of Nippur at the time of Ine-Sin, Bur-Sin II and Gimil (Kât)-Sin[6] in particular. All the

---

[1] Cf. Heuzey, *Revue d'Assyriologie* III, pp. 65–68. The description of this archive chamber excavated in Tello may find a place here: "Ces plaquettes de terre cuite, régulièrement superposées sur cinq ou six rangs d'épaisseur, remplissaient des galeries étroites, se coupant à angle droit, construites en briques crus et garnies des deux côtés de banquettes, sur lesquelles s'étendaient d'autre couches de semblables monuments. Les galeries formaient deux groupes distincts, mais voisins l'un de l'autre."

[2] The thievish Arabs seem to have scattered their rich harvest everywhere. So far, I have examined about 2000 of these tablets myself. But not less than c. 10,000 have been offered to me for sale by dealers of Asia, Europe and America within the last year. They all come from Tello. Cf. Hilprecht, *Recent Research in Bible Lands*, p. 80.

[3] Cf. Part I, pp. 27 f. and above, p. 230, note 1.

[4] For the proof of this statement cf. below.

[5] Cf. Pl. 13, No. 21, and Part I, "Table of Contents," p. 49. Bur-Sin II repeated only what had been done by Sargon I long before. Cf. Part I, "Table of Contents," p. 47 (No. 1), and below.

[6] That Gimil-Sin was the direct successor of Bur-Sin II follows from Pl. 58, No. 127, and that Ine-Sin was the immediate predecessor of Bur-Sin was inferred by Scheil from a contract tablet (*Recueil* XVII, p. 38, note 3). The mention of the devastation of Shashru on this Tello tablet is only of secondary importance in itself, as the same event

three kings mentioned devoted their attention to the interests of Inlil and Ninlil and other gods worshiped in Nippur, as we learn from excavated bricks and door-sockets (Pl. 12 f.),[1] from two chronological lists (Pl. 55, No. 125, and Pl. 58, No. 127),[2] and from the large number of dated contracts discovered in Tello, Nuffar and other Babylonian mounds.[3] That the country as a whole was quiet and enjoyed peace and prosperity under their government, is evident from the many business contracts executed everywhere in Babylonia and from certain statements contained in them. The constant references to successful expeditions carried on by Ine-Sin against the countries of *Karhar*[ki], *Harshi*[ki], *Simurrum*[ki],[4] *Lulubu*[ki], *Anshan*[ki] and *Shashru*[ki],[5] by Bur-Sin II

occurred at other times (*e. g.*, in Bur-Sin's sixth year, Pl. 58, No. 127, Obv. 6). But the fact that this conquest is placed between Bur-Sin's accession to the throne and a very characteristic event at the close of Ine-Sin's government (cf. Pl. 55, No. 125, Rev. 18-21) settles the question. Ine-Sin ruled at least forty-one years, according to the chronological list on Pl. 55. As, however, a part of it is wanting, it will be safe to assign a reign of c. 50 years to him. Bur-Sin II ruled at least twelve years (Pl. 58, No. 127), and in all probability not more than sixteen to eighteen years. That the events mentioned on the two tablets are arranged chronologically, is beyond question. For (1) events which happened more than once are quoted in their consecutive order, but often separated from each other by other events which occurred between them. Cf. Pl. 55, Rev. 3 and 10 ; Rev. 4, 5 and 11, and especially Obv. 5 and Rev. 15 (between the two similar events lie twenty-eight years !). (2) In case a year was not characterized by an event prominent enough to give it its name, such a year is quoted as "joined to" or "following" the previous year in which a certain event took place (*ush-sa*). Cf. Pl. 55, Rev. 7-8, 11-12, 13-14, 16-17, 18-20. (3) As we expect in a list arranged chronologically, Pl. 58, No. 127, opens with "the year in which Bur-Sin became king." If the king accomplished something worth mentioning in the year of his accession, this deed was added. Cf. Pl. 58, No. 127, Rev. 4 : *Mu* dingir*Gimil-* dingir*Sin lugal Urum*ki*-ma-ge ma-da Za-ap-sha-li*ki *mu-ġul-a* "In the year when (Gimil-Sin became king and =) King Gimil-Sin brought evil upon the land of Zapshali."

[1] Cf. also Peters in *The American Journal of Archæology* X, p. 16 f.

[2] Cf. No. 125, Obv. 2, 4, 10, 17, 18 (Ine-Sin), No. 127, Obv. 3, Rev. 3 (Bur-Sin II).

[3] Cf. for the present Scheil in *Recueil* XVII, p. 37 f.

[4] On a tablet in Constantinople written at the time of Ine-Sin, we read the following date : *mu Simu-ur-ru-um*ki *Lu-lu bu*ki *ba-ġul.* From the fact that *Simurru* and *Lulubu* are here mentioned together, Scheil (*Recueil* XVII, p. 38) draws the conclusion that "Simuru se trouvait donc dans les mêmes parages que là où la stèle de Zohab fixe le pays de Lulubi." This assertion is by no means proven. The king may have conquered two countries far distant from each other in the same year. I call attention to Scheil's theory in order to prevent conclusions similar to those which for several years were drawn from the titles of Nebuchadrezzar I (col. I, 9-11 : *sha danna* mâtu *Lulubî ushamkitu ina kakki, kâshid* mâtu*Amurrî, shâlilu Kashshî*) and led to curious conceptions about the land *Amurrî* (cf. *e. g.* Eduard Meyer, *Geschichte des Alterthums*, p. 329, and especially Winckler, *Untersuchungen*, p. 37, note 2). Hommel's identification of *Simurru* with *Simyra* in Phenicia is by far more probable (*Aus der babylonischen Altertumskunde*, p. 9).

[5] Pl. 55, No. 125, Rev. 3 ; resp. Rev. 6, 10 ; resp. Rev. 4, 5, 11 ; resp. Scheil, *l. c.*, p. 37 (beginning); resp. Rev. 13 ; resp. Rev. 21. In connection with Anshan it may be mentioned that Scheil in *Recueil* XVII, p. 38 (especially note 6), translated Pl. 55, No. 125, Rev. 9 : *mu dumu-sal lugal pa te-si An-sha-an*ki*-ge ba-tug* by "année où la fille du roi **devint** patesi dans le pays d'Anshan." Notwithstanding that Hommel (*Aus der babylonischen Altertumskunde*, p. 9) and Sayce (in *The Academy* of Sept. 7, 1895, col. b) reproduce this translation, which grammatically is possible, I reject it on the ground that there is no evidence that in ancient Babylonia women were permitted to occupy the highest political or religious positions independently, and translate : "In the year when the patesi of Anshan **married** a daughter of the king (*tug = ahâzu*, "to take a wife, to marry," cf. Delitzsch, *Assyrisches Handwörterbuch*, p. 42).

against *Urbillum^{ki}*, *Shashru^{ki}* and *Rite-tar*(?)*hu^{ki}*,[1] and by Gimil (Kât)-Sin against *Zapshali^{ki}*,[2] testify to the same effect. Moreover, a number of other tablets which belong to members of the same dynasty, but cannot yet be referred to definite kings, mention *Kimash^{ki}*, *Humurti^{ki}* and *Huhu*(*nu*)*ru^{ki}* [3] as devastated or invaded by Babylonian armies.[4] Several of these cities and districts were situated on the east side of the Tigris and must be sought in Elam and its neighboring countries. We begin now to understand why the Elamites soon afterwards when they invaded Babylonia made such a terrible havoc of the temples and cities of their enemies; they simply retaliated and took revenge for their own former losses and defeats.

4. When the Cassite kings conquered Babylonia, the site of the ancient archive chamber was long forgotten and buried under a thick layer of *débris*. Their own store-room, in which all the votive objects published on Pls. 18–27 and Pls. 60 f., Nos. 133–142, were discovered, was situated at the edge of a branch of the Shatt-en-Nil outside of the great S.-E. wall of the temple of Bêl.[5] The destruction of the archive under discussion must therefore have taken place between the overthrow of the second

---

[1] Pl. 58, No. 127, Obv. 2 ; resp. Obv. 6 ; resp. Obv. 7.

[2] Pl. 58, No. 127, Rev. 4.

[3] Cf. Scheil, *l. c.*, p. 38. The city of *Marhashi* (in N. Syria, according to Hommel, *l. c.*, p 9) is mentioned in connection with a daughter of Ine-Sin on Pl. 55, No. 125, Obv. 14.

[4] In view of all these facts above mentioned, Hommel will doubtless change his view (that the kings of the second dynasty of Ur "were apparently confined to this city, as they did not possess Sumer and also lost Akkad"). That they were not confined to Ur, but possessed the whole south is proven by their buildings in Eridu (I. *R.* 3, No. XII, 1, 2) and in *Nippur* (cf. also the statements of the two chronological lists). If Winckler's theory as to the seat of the *sharrut kibrat irbitti* was generally accepted (Hommel apparently does not accept it), the second dynasty of Ur by this very title would also have claimed N. Babylonia. Whatsoever our position may be as to the meaning of this and other titles, as a matter of fact, the kings of the second dynasty of Ur possessed the south of Babylonia, and it is impossible to believe that kings who were the lords of S. Babylonia and conquered parts of Arabia, Syria, Elam and other districts between the four natural boundaries defined in Part I, p. 25, note 4, and who doubtless in consequence of their conquests assumed the proud title "king of the four quarters of the world," should not have been in the possession of all Babylonia (the case of Gudea is entirely different). The kings of the second dynasty of Ur changed the title of their predecessors, not because they had lost Sumer and Akkad, but because they owned more than the old title indicated. The title of Sumer and Akkad—as I understand its meaning—is practically contained in that of "king of the four quarters of the world" (Part I, pp. 24 f.), and the kings of the second dynasty of Ur dropped it therefore for the same reason as Dungi, when he assumed the title *shar kibrat arba'im* (Z. *A*, III, p. 94). As to the meanings of the different titles, Hommel (whose latest opinion is briefly stated in *Aus der babylonischen Altertumskunde*, p. 8) and I agree entirely, differing from Winckler especially in his interpretation of *shar kibrat arba'im* and *shar mâtuShumeri u Akkadi* in the oldest Babylonian inscriptions down to Hammurabi. Notwithstanding that, or rather because I read and studied his *Altorientalische Forschungen* III, pp. 201–248, and all his previous papers on the same subject **sine ira et studio** again and again, I have been unable to convince myself of the correctness of his views. Tiele (*Z. A.*, VII, p. 368), Lehmann (*Shamashshumukin*, pp. 68 ff.), Hommel (*l. c.*) and I apparently reached similar conclusions on this important question.

[5] Cf. Part I, "Table of Contents," p. 48 (Pl. 8, No. 15). Cf. also Peters in *The American Journal of Archæology* X, p. 15.

dynasty of Ur and the beginning of the Cassite rule in Babylonia. The history of the temple of Bêl during this period is enveloped in absolute darkness. No single monument of the members of the so-called first and second Babylonian dynasties has yet been excavated in Nuffar. Apparently our temple did not occupy a very prominent place during their government. And how could it be otherwise? Their rule marks the period of transition from the ancient central cult of Bêl in Nippur to the new rising cult of Marduk in Babylon. Bêl had to die that Marduk might live and take his place in the religious life of the united country. Even the brief renaissance of the venerable cult of "the father of the gods" under the Cassite sway did not last very long. It ceased again as soon as the national uprising under the dynasty of Pashe led to the overthrow of the foreign invaders, who had extolled the cult of Bêl at the expense of Marduk in Babylon,[1] and to the restoration of Semitic power and influence in Babylonia, until under the Assyrian kings Esarhaddon and Ashurbânapal a last attempt was made to revive the much neglected temple service in the sanctuary of Nippur.

5. The breaking and scattering of the vases point to a foreign invasion and to a period of great political disturbance in the country. No Babylonian despot, however ill-disposed toward an ancient cult, and however unscrupulous in the means taken to suppress it, would have dared to commit such an outrage against the sacred property of the temple of Bêl. In all probability therefore the ancient archive chamber of the temple was ransacked and destroyed at the time of the Elamitic invasion (c. 2285 B.C.), when Kudur-Nankhundi and his hordes laid hands on the temples of Shumer and Akkad. That which in the eyes of these national enemies of Babylonia appeared most valuable among its contents was carried to Susa[2] and other places; what did not find favor with them was smashed and scattered on the temple court adjoining the storehouse. From the remotest time until then apparently most gifts had been scrupulously preserved and handed down from generation to generation. Only those movable objects which broke accidentally in the regular service, or which purposely were buried in connection with religious rites, may be looked for in the lowest strata of Ekur.

AGE OF THE INSCRIBED MONUMENTS.

Having explained why the most ancient documents so far excavated in Nuffar were found in pieces above the platform of Ur-Gur's ziggurrat, I now proceed to determine the general age of these antiquities and their relation to the inscriptions of Sargon I.

[1] Cf. Part I, pp. 30 f.
[2] Cf. Part I, p. 31.

The inscriptions Nos. 86–112 have many palæographic features in common and doubt-less belong to the same general period, the precise extent of which cannot be given. Two groups, however, may be clearly distinguished within it, differing from each other principally in the forms used for *mu* (Brünnow, *List* 1222) and *dam* (*ibid.*, 11105). Instead of the two familiar Old Babylonian characters, in *mu* the two pairs of parallel lines found at or near the middle of the horizontal line, sometimes cross each other (Nos. 92, 5; 98, 3; 99, 4; 101, 3, etc.), while *dam* occasionally has a curved or straight line between the two elements of which it is composed (No. 111, 3 and 6; No. 98, 2 and 5; cf. No. 94, 3).[1] This peculiar form of *dam* has so far not been met with outside of a very limited number of inscriptions from Nippur; that of *mu* occurs also on the barrel cylinder of Urukagina,[2] although in a more developed stage. Whenever one of these characters has its peculiar form in an inscription of Nippur, the other, if accidentally occurring in the same inscription, also has its peculiar form as described above (cf. No. 94, 3 and 4; No. 98, 2 (5) and 3; No. 111, 3 and 6). The two char-acters represent therefore the same period in the history of cuneiform writing, to the end of which the cylinder of Urukagina also belongs. This period has not yet been definitely fixed. As various historical considerations seemed unfavorable to placing this ruler after the other kings of Shirpurla, Jensen provisionally placed him before them;[3] Heuzey was less positive;[4] Hommel[5] and Winckler[6] regarded him as later, while Mas-pero, without hesitation, but without giving any reasons, made him "the first in date of the kings of Lagash."[7] Aside from the reasons given by Jensen, and a few simi-lar arguments which could be brought forth in favor of his theory, the following palæo-graphic evidence proves the chronological arrangement of Jensen and Maspero to be correct:

1. The peculiar form of *mu* occurs in inscriptions from Nippur which, if deter-

---

[1] This short line, about the significance of which I refer to my greater work, *Geschichte und System der Keilschrift*, was originally curved, became then straight and was later placed at the end of the character (No. 93, 6; 96, 4; 113, 12), finally developing into a full-sized wedge (De Sarzec, *Découvertes en Chaldée*, Pl. 26, No. 1, col. II, 1; Heuzey in *Revue d'Assyriologie* II, p. 79, No. 1, 13 [a duplicate of this inscription is in M. I. O., Constantinople], and the present work, No. 123, Obverse, 1). Sometimes this line is entirely omitted (No. 112, 6).

[2] De Sarzec, *l. c.*, Pl. 32, col. I, 7; col. II, 1, 4, 12; col. III, 3, 7. The form of *mu* is more developed in Uruka-gina's inscription, indicating that the latter is somewhat later than the corresponding Nippur texts. On the other monuments of Urukagina the regular Old Babylonian form is used exclusively.

[3] In Schrader's *Keilinschriftliche Bibliothek*, Vol. III, Part 1, p. 8.

[4] Formerly he regarded him as decidedly later than the other kings of Lagash (in De Sarzec, *Découvertes en Chal-dée*, pp. 110, 112). More recently he expressed himself as doubtful: "Il en résulte que le roi Ourou-ka-ghi-na doit être tenu, soit pour appartenir à une dynastie antérieure à celle du roi Our-Nina, soit pour avoir, après l'apparition des premiers patési, relevé le titre royal à Sirpourla" (*Revue d'Assyriologie* II, p. 84).

[5] *Geschichte Babyloniens und Assyriens*, pp. 290f.

[6] *Geschichte Babyloniens und Assyriens*, p. 41.

[7] *The Dawn of Civilization*, p. 604.

mined by the character of *dam* alone, must be classified as older than the royal inscriptions of Tello.

2. The form of *mu* employed in Urukagina's cylinder does not occur in any other inscription of Tello. The cylinders are therefore to be regarded as older than the other monuments, if it can be shown that this peculiar form of *mu* represents a more ancient stage of writing[1] and did not originate from an accidental prolongation of certain lines in *mu* by a careless scribe.[2]

3. The very pronounced forms cut in stone vases (as, *e. g.*, found in No. 98, 3; 101, 4; 92, 5, and first of all in No. 94, 4) force us to eliminate the element of accident. But, besides, it can be proved by an analysis of the character *mu* itself that the regular Old Babylonian sign is only a later historical development of a more ancient form. The correct interpretation of the original picture will, at the same time, enable us to catch an interesting glimpse of certain prehistoric conditions in ancient Shumer. According to Houghton,[3] a close relation exists between the character for *mu* and *ḥu* (Brünnow, *l. c.*, 2044) and the first part of the character for *nam* (*ibid.*, 2087). I trust no Assyriologist of recent date has ever taken this attempt at solving a palæographic problem very seriously. The sign for *nam* has no connection with the other two characters and is no compound ideogram, but, in its original form, represents a flying bird with a long neck.[4] Since in Babylonia, as in other countries of the ancient world, the future was foretold by observing the flight of birds, this picture became the regular ideogram for "fate, destiny" (*shimtu*) in Assyrian. The original picture for *mu*, on the other hand, is no bird, but an arrow whose head formerly pointed downward, and whose cane shaft bears the same primitive marks or symbols of crossed lines as are characteristic of the most ancient form of arrow used in the religious ceremonies of the North American Indians.[5] As the shaft was represented by a single line in Baby-

[1] This argument is conclusive, as the theory, according to which later writers occasionally imitate older forms of cuneiform (or linear) characters, in the sense generally understood by Assyriologists, is without any foundation and against all the known facts of Babylonian palæography. Cf. my remarks in Part I, pp. 12f.

[2] Jensen's hesitation, so far as founded upon the form of the character *ka*, can be abandoned, as the form of this character is surely far older than Gudea.

[3] In the *Transactions of the Society of Biblical Archæology* VI, pp. 464f.

[4] This fact becomes evident from a study of the oldest forms in the inscriptions of Tello and Nippur. The original picture is still found on the most ancient Babylonian document in existence, unfortunately scarcely known among Assyriologists. It is (or was) in the possession of Dr. A. Blau and was published by Dr. W. Hayes Ward in the *Proceedings of the American Oriental Society*, October, 1885. The bird represented is therefore no "swallow" (Hommel, *Sumerische Lesestücke*, p. 6, No. 67), but a large bird with a long neck, such as a goose or a similar **water** bird found on the Babylonian swamps. Later our picture was also used as the ideogram for "swallow," designating her as the flying bird *par excellence*, as the bird nearly always in motion when seen at day time.

[5] As I learned through the courtesy of Mr. Frank Hamilton Cushing of the Bureau of American Ethnology in the Smithsonian Institution at Washington. After a correspondence on this subject it became evident that we had

lonian writing, the original mark carved upon its surface had to be drawn across it. Instead of ⟶, we find, therefore, ⟶, from which, by shortening the crossed lines, the regular form ⟶ developed at a later time. The correctness of this explanation is assured by the otherwise inexplicable absence of an ideogram for *uṣṣu*, "arrow," in Assyrian. For it is impossible to conceive that a people using the bow in their system of writing should have altogether excluded the arrow, which played such a conspicuous *rôle* in the daily life and religious ceremonies of ancient nations in general. But how is it to be explained that our ideogram does not mean "arrow" at all, but signifies "name?" Just as the picture of a flying bird in writing proper was used exclusively with reference to its religious significance, in order to express the abstract idea of "fate, destiny," so the arrow with the marks or symbols of ownership (originally two crossing lines [1]) carved on the shaft became the regular ideogram for "personality" or "name." The same association of ideas led to exactly the same symbolism and usage among the North American Indians, with whom "the arrow" is the symbol of personality.[2] It becomes now very evident that the Babylonian seal-cylinder, with its peculiar shape and use, has developed out of the hollow[3] shaft of an arrow marked with symbols and figures, and is but a continuation and elaboration in a more artistic form of an ancient primitive idea.

From palæographic and other considerations it is therefore certain that Urukagina lived before the ancient kings of Shirpurla, while the inscriptions published in the present work as Nos. 90, 91, 92, 94, 98, 99, 101, 111 are still older than Urukagina. The interval between him and the following rulers of Tello who style themselves "kings" cannot have been very great, however. They all show so many palæographic features in common that they must be classified as an inseparable group. To the

both reached the same conclusions as to the oldest form and significance of the arrow in picture writing by pursuing entirely different lines of research. My arguments, corroborated by Mr. Cushing's own investigations and long residence among tribes which still practice many of the ancient primitive rites and customs, become therefore conclusive in regard to the original form of the character *mu*. I quote from Mr. Cushing's letter the interesting fact that the above drawn arrow with two pairs of crossing lines on its shaft is called by the Zuñi *a'thlua* "speeder (commander) of all" (namely, of all the other arrows used in their religious ceremonies). A treatise on the ceremonial use of the arrow among the Indians, by Mr. Cushing, is in press.

[1] Still used with the same significance in Europe and America by persons who cannot write, if they have to affix their names to legal documents. The crossed lines on the Indian arrows have a deep religious significance, according to Cushing.

[2] Cf. on this whole subject Culin, *Korean Games*, pp. XXIf. To Prof. Dr. Brinton and Mr. Stuart Culin I am indebted for recent information on this subject.

[3] Because made of bulrushes, growing abundantly along the marshes and canals of lower Babylonia.

same age doubtless belong most, if not all, of the other inscriptions published on Pls. 36–47 (No. 112). I shall prove my theory in detail by the following arguments:

I. Palæographically they exhibit most important points of contact with Urukagina, Ur-Ninâ, Edingiranagin, Enanatuma I, Entemena, Enanatuma II, especially with the first three mentioned.

a. Characteristic signs are identical in these Nippur and Tello inscriptions. Cf., *e. g.,* **gish**, No. 87, col. I, 10, col. II, 37, No. 110, 4 f. e., with the same sign in the texts of Ur-Ninâ and Edingiranagin;[1] **ban**, No. 87, col. I, 10, col. II, 37 (cf. No. 102, 2) with the same sign in the texts of Edingiranagin; **a**, No. 86, 8 (Var.), 1 f. e., No. 87, *passim;* No. 96, 2; No. 104, 3; 106, 4; 110, 8 f. e., 112, 7, with the sign used by Ur-Ninâ, Edingiranagin, Enanatuma I, Entemena (cf. also the present work, No. 115, col. I, 7, col. II, 1, 2, etc.); **shú**, No. 87, col. III, 34 (and Var.) with Urukagina, Edingiranagin; **da**, No. 86, 7, No. 87, col. I, 19, col. II, 18, 20, 29, etc., with the sign used by Ur-Ninâ, Edingiranagin, Entemena; **à** (ID), No. 87, col. II, 41 (Var.) with Entemena (No. 115, col. I, 5); **ta**, No. 87, col. I, 46, col. II, 4, 12, with the same sign used by Urukagina, Ur-Ninâ, Edingiranagin, Entemena; **mà**, No. 88, col. III, 2, with the same used by Urukagina, Endigiranagin;[2] **ma**, No. 87, col. II, 40 ff., with the same sign used by Urukagina, Edingiranagin; and many other characters.

b. The script is almost entirely linear like that of Urukagina,[3] Ur-Ninâ and Edingiranagin.

c. They show certain peculiarities in the script, which so far have been observed only in the most ancient texts of Tello: (1) Lines of linear signs running parallel to a separating line (marking columns and other divisions) frequently fall together with this latter so that the character now appears attached to the separating line above, below, to the right or left. Sometimes characters are thus attached to two separating lines at the same time. Cf. No. 87, col. I, 5 (*ma*), 12 (*ka*), col. II, 9 (*shu*), 17 (*la*), 29 (*li*), col. III, 36 (*ur*), No. 106, 2 (*nin*), and many others written on different fragments of No. 87.[4] (2) In accordance with this principle two or more characters

---

[1] In these quotations, as a rule, I shall abstain from giving the exact passages, as I expect that everybody who examines my arguments has made himself familiar with the palæography and contents of the most ancient inscriptions of Tello before, and to those who have not done so, I do not intend to give introductory lessons in the limited number of pages here at my disposal, in fact for those I do not write.

[2] Also used by Narâm-Sin, cf. No. 120, col. II, 4.

[3] Except of course his barrel cylinder, which has cuneiform characters, as it was inscribed with a stylus.

[4] For this palæographic peculiarity in the inscriptions of Tello, cf. **Urukagina** (De Sarzec, *Découvertes*, Pl. 32, col. II, 9, 10, col. III, 2, 5, col. IV, 3, 9, col. V, 2, 4); **Ur-Ninâ** (De Sarzec, *l. c.,* Pl. 2, No. 2, col. I, 1, 3, *Revue d'Assyriologie* II, p. 84, 3 and 4; p. 147, col. I, 3, 5, col. III, 3, 6, col. IV, 3, 5); **Edingiranagin** (De Sarzec, *l. c.,* Pl. 4, Frag. A, col. I, 6, col. II, 3, 4, 5, 10, etc.; Pl. 31, No. 2, col. I, 1–4, 6, col. II, 1–3, 5, etc.); **Enanatuma I** (*Revue*

standing in close proximity to each other frequently enter into a combination, forming so-called ligatures.[1] Cf. No. 86, 5 Var. (*mà-na*), 8 (*lab-ba*, cf. also Variants), 15 Var. (*ki-gub*) ; Part I, Pl. 14, 2 (*du-du*) ; No. 87, col. II, 9 (*ma-shu*), 20 Var. (*da-gà*), 34 (*ki-ag*), 45 (*da-gi*, cf. Var. *gi-gi*),[2] col. III, 21 (*ba-daġ*),[3] 34 (PA [first half of the character *sib*][⁴-*gal*]; No. 93, 7 (*Shul-pa*);[5] No. 94, 1 (*Nin-din-dug* (?)) ;[6] No. 98, 2 (*dam-dumu*); No. 111, 6 (*na-dú*).[7] On the monuments of Tello this tendency to unite two characters into one is almost entirely confined to the inscriptions of Ur-Ninâ.[8] The best illustration is afforded by the writing of the name of his son, Ninâ-shu-banda. The four signs which compose the name are contracted into one large sign, the earliest example of a regular monogram in the history of writing (De Sarzec, *l. c.*, Pl. 2^(bis), No. 1). A number of signs which occurred always[9] in the same

---

d'Assyriologie III, p. 31, 1-5, 9, 11, 14 f.); **Entemena** (De Sarzec, *l. c.*, Pl. 5, Nos. 2, 4 and 5 ; Pl. 31, No. 3, col. I, 2, 4, 5, col. II, 3 ff ; *Revue d'Assyriologie* II, p. 148, col. I, 1-6, etc.) ; **Enanatuma II** (De Sarzec, *l. c.*, Pl. 6, No. 4, 2-5, 7 f.) For other examples of Entemena's text in the present work, cf. Nos. 115-117. Apparently Dr. Jastrow had not seen a Tello inscription when he wrote his remark in *Z. A.* VIII, p. 217.

[1] In a limited measure the same peculiarity occurs in several Assyrian inscriptions, c. 3000 years later. Cf., *e. g.*, *i-na*, in the inscription of Tiglathpileser I (I *R.*, 9 ff.), *ina pa*, *Salm. Obel.*, l. 160, 176 (Hilprecht, *Assyriaca*, p. 27, note),etc.

[2] Col. II, 43. *ki-nin Unug*^(ki)*-gà*, 44. *ganam-ġad-shakir-a-dim*, 45. *shig mu-da-gi-gi*. The last character in l. 38, which remained unidentified for such a long time (cf. Amiaud et Méchineau, *Tableau Comparé*, No. 122, Jensen in Schrader's *K. B.* III, part 1, p. 16, note 4 ; Scheil in *Recueil* XV, p. 63 ; Hommel, *Sumerische Lesestücke*, p. 32, No. 376) is identical with Brünnow, *List* 5410. It has in the ancient inscriptions the two values *gà* and *mà* (for the latter cf., *e. g.*, No. 87, col. II, 19 (*kalam-mà*), 29 (*Urumk*^i *-mà*)). On Pl. 50, col. II, 4, read NA-GA = *ishkun* (and col. III, 4 f., KI-GAL (= *kigalla*) *ish pu-uk*, against Scheil in *Recueil* XV, 62 f.).

[3] Col. III, 19. *nam-ti-mu*, 20. *nam-ti*, 21. *ġa-ba-daġ-ġi*—"unto my life he may add life."

[4] *PA-gal LU sag gud*, read *sib* (*PA-LU sag-guda-gal*, "the shepherd having the head of an ox" = "the ox-headed shepherd," a synonym of king, according to Jensen.

[5] On the god *Shul pa-ud-du*, cf. Jensen, *Kosmologie*, pp. 126 f., and in Schrader's *K. B.*, III, part I, p. 65, note 11 (*Umun-pauddu*). Oppert read *Dun-pa-e*.

[6] "The goddess who destroys life," an ideogram of Bau or Gula (Brünnow, *List* 11084, cf. III *R.*, 41, col. II, 29-31 ; III *R.*, 43, col. IV, 15-18, and the present work, Pl. 67, col. III, 1-5). The same deity is mentioned No. 93, 1, No. 106, 1, No. 111, 1. On the value of *dug* cf. Hommel, *Sumerische Lesestücke*, p. 5, No. 53, and p. 12, No. 145.

[7] Cf. No. 99, 5.

[8] Cf. *Revue d'Assyriologie* II, p. 147, col. III, 6 and 7, col. V, 1, 3, 6.

[9] Cf. No. 87, col. I, 5, 40, 42, etc. The linear sign is composed of *e* (canal) + *gi* (reed) and originally denotes a piece of land intersected by canals and covered with reeds (cf. No. 87, col. III, 29). The land *par excellence* with these two characteristic features was to the Babylonians their own country, which therefore was called by the oldest inhabitants *Ki* + *e* + *gi* = *Kengi*, "the land of canals and reeds." From this correct etymology of *Kengi* and its use in the earliest texts (*bar bar Kengi*, No. 87, col. II, 21, and *Enshagsagana en Kengi*, No. 90, 3) it follows that the name does not signify "low-lands" or "Tiefebene" in general in the ancient inscriptions, which alone have to decide its meaning (against Winckler in *Mitteilungen des Akademisch-Orientalistischen Vereins zu Berlin*, 1887, p. 12), but that it is the geographical designation of a well-defined district, **Babylonia proper**. As, however, Babylonia and low-lands are equivalent ideas, *Kengi* could also be used in a wider sense for "low-lands" (*mâtu*) in general.

combination and served to express but one idea or object, were regularly contracted at this early time and became compound ideograms, e. g., *kalama* " country," *gishdin*[1] " wine," etc. (3) Lines of linear signs which run parallel to a separating line are often omitted, even if the sign is not directly connected with this latter. Cf. No.

[1] The peculiar way in which it is written in the oldest inscriptions of Tello, leaves no doubt as to its composition (*gish* + *din*). The analysis of this ideogram by Pinches (*Sign List*, No. 76 a = *kash* + *din*), accepted by Delitzsch (*Assyrisches Handwörterbuch*, p. 354), Jensen (in Schrader's *K. B.* III, part 1, p. 27, note 6), Hommel (*Sumerische Lesestücke*, No. 180) and others, must therefore be abandoned. For examples cf. Edingiranagin's inscription unearthed in London (*Proc. Soc. Bibl. Arch.*, Nov. 1890), col. IV, 3, 7, col. V, 3 : *gishdin zu-zu-a;* or Gudea D (De Sarzec, *l. c.*, Pl. 9) : 6. *Ma-gan*ki, 7. *Me-lug-ga*ki, 8. *Gu-bi*ki, 9. *kur Ni tug*ki, 10. *gu gish mu-na gal la-a-an*, 11. *ma gishru-a gishdin* (*sic!*), 12. *Shir-pur-la*ki-*shu*, 13. *mu-n i-tum*—" Magan, Meluḫa, Gubi, Dilmun, each (*an*) of which possesses every kind of tree, brought a ship (laden) with timber and wine to Shirpurla." Jensen's question (in Schrader's *K. B.* III, part 1, p. 13, note 12), as to what Amiaud may have read in Ur-Ninâ's inscription (De Sarzec, *l. c.*, Pl. 2, No. 1, col. IV, 1-3, which Jensen left untranslated) is answered by referring him to the Gudea passage just translated, and to *Revue d'Assyriologie* II, p. 147, col. V, 3-6, together with De Sarzec, *l. c.*, Pl. 2bis, No. 1 (lower section, characters standing immediately before the king). Amiaud, however (in *Records of the Past*[2] I, p. 65), as well as Oppert (in *Revue d'Assyriologie* II, p. 147) and Heuzey (in *Revue d'Assyriologie* III, p. 16, and *Découvertes en Chaldée*, p. 170) wrongly read *gish din* (notwithstanding the passage from Gudea just quoted, lines 6 and 10, where the two respective characters are very different from each other !) as *gan* (*kan*) finding the name of Magan in the first line. The passage reads rather : 1. *ma gishdin*, 2. *kura-ta*, 3. *gu gish gal*, 4. *mu-tum* (?)—"a ship (laden) with wine he brought from the country which possesses every kind of tree." We are now enabled to understand the full significance of Ur-Ninâ's perforated bas-relief (De Sarzec, *l. c.*, Pl. 2bis) which remained obscure to Heuzey in his treatise mentioned below. These bas-reliefs and incised slabs (cf. the present work, Pl. XVI, Nos. 37 f.) did not serve "a maintenir dressés, sur des autels ou sur des massifs de briques, divers engins consacrés aux dieux et particulièrement des masses d'armes votives" (Heuzey, *Les Armoiries Chaldéennes de Sirpourla*, pp. 11 f., cf. pp. 6 f.). For they would have been too small and weak for such a purpose. The true facts are rather these : (1) They accompanied donations of any kind made to the temple. But while such donations were consumed in the interest of the temple service (cf. Hilprecht, *Z. A.* VIII, p. 191 f.) or decayed in time (buildings) or died (slaves), etc., these tablets were preserved in the temple as lasting memorials to their munificent donors and served at the same time to induce other worshipers to similar acts of piety. (2) The hole in the middle of the tablets served to fasten it, by the aid of a nail, in the wall or floor of the temple, possibly on the altar itself. (3) The scenes, objects and inscriptions on these tablets generally illustrate and describe the person and work of the donor in relation to his deity. Ur-Ninâ's more elaborate votive tablets (of which the smaller is only an excerpt, cf. De Sarzec, *l. c.*, Pl. 2bis, pp. 168-173), accordingly represent two sides of the king's work undertaken in the service of his god. In the upper section he has the *dupshig* (= *dupshikhu*), the symbol of masons, upon his head (exactly as Nabopolasser describes himself in the present work, Pl. 33, col. II, 57 ff.), and is surrounded by his children and page (*Da-ni ta* " at his side " = " page," not "in his hand,"—Oppert in *Revue d'Assyriologie* III, p. 16, note 1). This picture illustrates the accompanying statement : " Ur-Ninâ, king of Shirpurla, son of Nigalnigin, built the temple of Ningirsu, built the *abzu-banda* (cf. Jensen in *K. B.* III, part 1, p. 13, note ††), built the temple of Ninâ." In the lower section the same king, seated and surrounded by his children and his chief butler (*Sag-antug* " he is the chief "), offers a libation of wine. This picture illustrates the words standing below the cup, "a ship of wine he brought from the country which possesses every kind of tree." The inscription of the bas-relief published by Heuzey in *Les Armoiries Chaldéennes de Sirpourla* reads : 1. *Lag* (DU-DU = *abâlu* "to bring," *nazâzu* "to set up "), 2. *sanga* (Brünnow, *List* 5980) *may*, 3. dingir *Nin gir su-ka*, 4. dingir *Nin-gir-su*, 5. *E-ninnû-ra*, 6. *lag*, 7. *sanga* (cf. the present work, No. 87, col. I, 30, and No. 113 ,3) dingir *Nin gir-su-ka-ge*, 8. . . . . ki *ta*, 9. *mu-na-ta-ud-du*, 10. GAG + GISH (not *gisal*, Hommel, *Sum. Lesest.*, No. 205) *ura-shu*, 11. *mu-na-gim* —" Gift of the high-priest of Ningirsu to Ningirsu of the temple Eninnû. The gift of the priest of Ningirsu he brought from . . . . and worked it into a . . . ."

86, 3 Var. (*ra*), 4 Var. (*li*), 5 Var. (*na*); No. 87, col. I, 4 (*Unug*), 14 and 20 Var. (*dingir*), 19 Var., col. II, 37 Var., 45, III, 34 Var. (*da*), 40 Var. (*kalama*); col. II, 31 Var. (*gim*); col. III, 2 (*um*), 23, 41 Var. (*à*), 29 (*mà*), 37 Var. (*nam*), etc. Outside of the Nippur texts this peculiarity is almost confined[1] to the inscriptions of Ur-Ninâ. Cf., *e. g.*, De Sarzec, *l. c.*, Pl. 2 [bis], No. 2, upper section (*da* in the name of *Ab-da*), *ibid.* (*Ur* in the name of Ur-Ninâ), *Revue d'Assyriologie* II, p. 147, col. V, 4.

II. The palæographic evidence brought forth is conclusive. Nos. 86, 87 and the other texts referred to above, show all the characteristic features of the inscriptions of Urukagina, Ur-Ninâ and Edingiranagin. But besides they exhibit a number of palæographic peculiarities which are altogether absent from the inscriptions of Tello, and must be regarded as characteristic features of an earlier stage of writing. They will be treated in full at another place.[2] I confine myself here to a brief statement of the following fact. A number of signs have a form representing almost the original picture, others have at least a more original form than the inscriptions from Tello, even those of Urukagina not excepted. Cf. *sum* (No. 87, col. I, 17, the ear of a corn, cf. also l. 45), *gi* (*ibid.*, col. I, 3, a reed, bulrush)[3], *à* (*ibid.*, col. I, 31 in *egi-a*, a tattooed forearm with hand),[4] *bar* (*ibid.*, col. II, 21; No. 98, 4 (the skin of an animal or) a coarse rug),[5] *lah* (*ibid.*, col. 1, 21, water poured out, therefore, "to wash"),[6] *ra* (*ibid.*,

---

[1] One example is found in a text of Entemena (*ne*, cf. *Revue d'Assyriologie* II, p. 149, col. IV, 2). The way in which *Ur* is written in the name of Urukagina (De Sarzec, *l. c.*, Pl. 32, col. I, 1), furnishes the key to the origin of this peculiarity. For details on this subject I refer to my *Geschichte und System der Keilschrift*, which has been in preparation for the last nine years.

[2] In advance I warn Assyriologists not to regard a fourth palæographic peculiarity (so far confined to these Nippur texts) as a mistake of the scribes: (4) If two linear signs which are to be connected grammatically stand close together in writing, yet without touching each other, frequently one line of the second running parallel to a line in the first is omitted entirely and has to be supplemented from the first sign. Cf. No. 87, col. III, 37: *la-ni* (*sic!*), 39: *aga-ni* (*sic!*), 40 Var.: *mu-na* (*sic!*); No. 103, 3: *mà-na*.(*sic!*).

[3] In order to obtain a clear conception of the original picture, this sign must not be turned to the left (as Houghton, *l. c.*, p. 473, and others did). For it is a law in cuneiform writing "that the characters are all and always reversed in the same way; what (originally) was the right-hand side became (later) the top" (Bertin, *l. c.*, p. 6). The triangle on the left of our picture does not represent the lower end of the stem of a reed, but rather its top or cob. Cf. the corresponding pictures on the Assyrian monuments published in Layard, *The Monuments of Nineveh*, Second Series, *e. g.*, Pl. 12, No. 1 (reproduced by Maspero in *The Dawn of Civilization*, p. 561).

[4] The crossed lines do not represent "an ornamented sleeve" (Bertin, *l. c.*, p. 9), but marks of tattooing (cf. Berger, "Rapport sur les tatouages Tunisiens," in *Revue d'Assyriologie* III, pp. 33–41). The cuneiform sign without these marks means "side" (*da*); with them, it denotes him who is at somebody's side for assistance; he who has the same marks of tattooing upon his arm, therefore has become his "brother." The sign for *shesh*, "brother," denotes a person as the second child of the same family, while the former expresses tribal relations represented by a common symbol.

[5] According to Oppert (*Expédition en Mésopotamie*, Tome II, p. 64) and Bertin (*l. c.*, p. 8) an altar. Impossible! It represents the skin of an animal or better a coarse rug spread upon the ground for persons of rank (and images of deities) to sit upon; in other words, it denotes the place of honor, in exact harmony with the custom prevailing in the tents of Arabia and Mesopotamia to-day. Lehmann (*Shamashshumukin*, p. 122) is therefore correct in giving

col. I, 37 Var., col. III, 15 Var., "canal" + "to fill" (si = horn), *i. e.*, "to irrigate"),[1] *lugal* (*ibid.*, col. I, 1–3, the sign shows the remnant of the original arm.[2] Cf. also the ideogram *zag* (*ibid.*, col. I, 3, 38, etc.), *gur* (*ibid.*, col. III, 42 Var.),[3] *Kish* (No. 92, 3; No 102, 3; 103, 4),[4] *ag* (No. 83, 11 and 14),[5] and many others for whose explanation I must refer to my *Geschichte und System der Keilschrift.*[6] All the stone inscriptions of Urukagina have the regular Old Babylonian sign for *mu*,[7] just as the Nippur texts here treated. On the other hand, the Nippur texts have a large number of far more original forms of signs than the Urukagina and Ur-Ninâ inscriptions published.[8] In view of these facts I can only draw one conclusion—that most of these Nippur texts are older than those of Urukagina.

III. Another important fact corroborates my determination of the age of these

---

to *bara(g)* the original meaning, "seat," instead of "chamber." This sign occurs frequently in the contracts of Nuffar (in a much more developed form) and was identified with *bar* by Scheil independently of me. Cf. *Recueil* XVII, p. 40d.

[6] *Suk(k)allu* denotes the servant (*gal*) who pours out (*su*) [namely water over his master's hands and feet]. A word with similar meaning (*zu*) is apparently contained in *zu-ab*, "ocean," which Hommel translated half correctly "house of water (?)," cf. *Sumerische Lesestücke*, No. 6. Originally *zu* and *su* had the same ideogram, which represents a vessel (cistern?) into which water flows. *Zu* means, therefore, "to flow into," or trans., "to pour into, to add," then figur., "to increase one's knowledge, to learn, to know." *Zu-ab* denotes "the house (abode) into which all the waters flow." *Sukkallu* may be translated "chamberlain" (Kämmerer), later it received a more general meaning.

[1] Oppert already recognized the general significance of the picture (*l. c.*, p. 64). But the exact analysis of the compound ideogram, which I discovered long before we excavated in Nuffar, remained obscure to him, Houghton, Sayce (*Transactions of Soc. Bibl. Arch.* VI, p. 475) and others. Cf. a very curious form, which is but a mutilated "*ra*," in col. I, 37, second Var.

[2] The two elements *lu* + *gal* appear separated in No. 83, 2 Var., 13 Var.; No. 104, 7; No. 105, 7.

[3] Successfully analyzed by Ball in *Proc. Soc. Bibl. Arch.* XV, p. 49. The line which continues beyond the head is, however, no continuation of the forearm, but represents the cushion between the head and the vessel upon which the latter rests. Originally the arm reached further to the rim of the vessel, as in the corresponding Egyptian hieroglyphics and as illustrated by Pl. XVI, No. 37, of the present work.

[4] It closely approaches the original picture explained by a Babylonian scribe on the famous fr. from Kuyunjik, col. III, 6 (*Trans. Soc. Bibl. Arch.* VI, p. 455).

[5] Cf. also the same sign on the very ancient monument preceding Urukagina's time (De Sarzec, *l. c.*, Pl. 1bis b., col. IV, 1).

[6] As I have to dispose of more urgent matters at present, some years may still pass before its publication.

[7] Only his barrel cylinder in clay exhibits traces of the older form for *mu*, as shown above.

[8] Nobody can object that a few characters in these Nippur inscriptions **seem** to show the beginning of wedge-writing and that a few others **seem** to have a later form. Lugalzaggisi presented c. 100 large inscribed vases, all apparently bearing the same long inscription here published, to Inlil of Nippur. Every stonecutter available was employed. Several of them understood but little of writing, and consequently some very ridiculous forms were produced. Cf., *e. g.*, col. II, 16 (second variant), *dug-a* (*sic!*), 29 (second variant) *da*, 39 (variants) *aga*, 42 *gur*, 44 (fourth variant) *ganam*, 45 *shig*, and others. In order to understand the enormous difficulties which I had to overcome in restoring this text, Assyriologists will bear this fact in mind.

inscriptions very strongly. In the inscriptions of Edingiranagin, or Edingiranatum,[1] the grandson of Ur-Ninâ, a city, generally transliterated as *Is-ban*[*ki*], plays a very important *rôle*. In fact the annihilation of the power of this city in S. Babylonia is the one prominent feature which characterizes his government, and to which (in connection with Erech, Ur and some other cities) the king refers again and again.[2] The most interesting object yet found in Tello, the so-called stele of vultures, was doubtless set up by this sovereign in commemoration of his great victory over *gish*BAN[*ki*].[3] However this may be, so much is certain that at some time previous to Edingiranagin, a foreign power whose centre was *gish*BAN[*ki*], had succeeded in invading and conquering a large portion, if not the whole, of Babylonia, Erech and Ur included. The same city of *gish*BAN[*ki*] is also mentioned in the long Nippur text No. 87, and here again it occurs in connection with Erech and Ur (and Larsam). We learn at the same time from this very important historical document that Lugalzaggisi, son of a certain Ukush "patesi of *gish*BAN[*ki*]"[1] (col. I, 3, 9, 10) had conquered all Babylonia and established an empire extending from the Persian Gulf to the Mediterranean Sea, in size therefore not inferior to that founded much later by Sargon I. This first "king of the world" (*lugal kalama*, col. I, 4, 36–41, col. III, 4) of whom Babylonian documents give us information, selected Erech as his capital, and by his great achievements raised *gish*BAN[*ki*], his native city, "to great power" (*à maǵ mu-um-gur*, col. II, 41f.). The two documents, Nippur, No. 87, and the stele of vultures from Tello, belong closely together and supplement each other, the one giving a *résumé* of the rise and height of the power and influence of *gish*BAN[*ki*], the other illustrating its downfall. The former must therefore antedate the monument of Edingiranagin. As doubtless some time elapsed between the rise and downfall of this foreign power; as, moreover, Shirpurla is not mentioned in Lugalzaggisi's inscription, apparently because it did not as yet exercise any political influence;[5] and finally as palæographically this inscription from Nippur shows more traces of originality than the texts of Urukagina and Ur-Ninâ, as

---

. [1] In view of De Sarzec, *l. c.*, Pl. 31, No. 2, col. III, 5 (*E-dingira-na-tum-mà* = "Brought into the house of his god " (by his parents after his birth).

[2] Cf. De Sarzec, *l. c.*, Pl. 3, Fragm. A, col. I, 5, 8, col. II, 4, 13, col. III, 5 ; Pl. 4, Fragm. A, col. II, 2, 11 ; Fragm. B, col. III, 3, col. V, 4 ; Pl. 31, No. 2, col. I, 6.

[3] For details cf. Heuzey's explanation of the figurative representations in his work, *Les Origines Orientales.* pp. 49–84, and in De Sarzec, *l. c.*, pp. 174–184. I agree with this scholar that the people whose defeat is illustrated on this monument belong to the city (and country) of *gish*BAN[*ki*] (De Sarzec, *l. c.*, pp. 183).

[4] This was the original reading of l. 10 ; the traces preserved on two fragments establish my text restoration of this line beyond doubt.

[5] The fragment of an inscribed object, apparently dedicated by a king of *gish*BAN[*ki*] to Ningirsu, was found in Tello (De Sarzec, *l. c.*, Pl. 5, No. 3, and p. 119). From the character used for "king" I draw the conclusion (with Heuzey) that the object belongs to a somewhat later period. Apparently *gish*BAN[*ki*] played a second important *rôle* in the Babylonian history.

stated above, we are justified in placing Lugalzaggisi before these two rulers of Shir-purla and in regarding most of the inscriptions published as Nos. 86–112 as older than the earliest royal inscriptions from Tello.[1] At any rate, they are not later than these.

A question of fundamental importance for our correct conception of the earliest phase of Babylonian history has been repeatedly discussed within the last ten years: In which relation did Sargon I (and Narâm-Sin) stand to the early kings of Tello? Did he antedate or succeed them? Winckler[2] and Maspero[3] expressed themselves decidedly in favor of the former view,[4] while Hommel,[5] Heuzey[6] and myself (Part I, p. 19),[7] with more or less emphasis placed Sargon I and his son after Ur-Ninâ and Edingiranagin I will now briefly give the definite proof of the validity of our theory.

1. The results of the exploration of the lowest strata of Ekur will have convinced. us that Babylonian civilization had a history antedating the kingdom of Sargon I by several thousand years. This pre-Sargonic period must have had a system of writing; for the earliest texts at our disposal, however closely approaching the original picture in a number of cases, presuppose an earlier stage of writing, such as is testified to have existed in Babylonia by the monument "Blau"[8] and by the famous fragments from Kuyunjik.[9] Pieces of inscribed objects unearthed below the Sargon level prove posi-tively that writing existed in Nippur long before Sargon I. It seems, therefore, at the very outset, impossible to believe that not one document antedating the highly devel-oped style of writing in Sargon's monuments should have been excavated in Nuffar or Tello. In fact, it would be altogether unreasonable to regard the inscriptions of Sargon and Narâm-Sin as the first written records of the ancient Babylonian civili-zation.

2. Everybody who has **studied** the earliest inscriptions of Babylonia from their originals, and has devoted that special pains to all the details of palæography, which

[1] The little fragment No. 107 cannot be referred to the time of Entemena, the only other ruler of Tello who, according to our present knowledge, presented an inscribed vase to Inlil. Perhaps it is the first indication of the rising of Shirpurla in the South and of the extending of its sphere of influence northward at the expense of gishBANki.

[2] *Untersuchungen*, p. 43 ; *Geschichte*, pp. 40f. (but cf. on the other side p. 42 !); *Altorientalische Forschungen* III, pp. 236ff.

[3] In *Recueil* XV, pp. 65f.; *The Dawn of Civilization*, p. 605. note 3 (end).

[4] Recently adopted by Rogers, *Outlines of the History of Early Babylonia*, Leipzig, 1895, p. 11, note 1 [but given up again after hearing my address, *Contributions to the History of Sargon I and His Predecessors*, before the Oriental Club of Philadelphia].

[5] *Zeitschrift für Keilschriftforschung* II, p. 182 ; *Geschichte Babyloniens und Assyriens*, p. 291.

[6] Cf., e. g., *Les Origines Orientales*, pp. 50, 84 ; *Revue d'Assyriologie* III, pp. 54, 57.

[7] Cf. also *Recent Research in Bible Lands*, pp. 66f.

[8] Called so for the sake of brevity. Cf. above, p. 249, note 4.

[9] Published by Houghton in *Trans. Soc. Bibl. Arch.*, p. 454, and reproduced in several other works.

I have a right to expect from those who criticise my statements on this subject, must necessarily come to the conclusion that a much longer period of development lies between Lugalzaggisi, Urukagina, Ur-Ninâ and Edingiranagin, on the one hand, and Sargon and Narâm-Sin, on the other, than between the latter and Ur-Ba'u Gudea, Ur-Gur, etc. It is surely remarkable that Monsieur Heuzey[1] and myself, who have devoted years of constant study to the palæography of the earliest original inscriptions of Babylonia, quite independently of each other, have reached exactly the same conclusions. It is out of regard for the view of those who do not accept Nabonidos' 3200 years as correct, that on palæographic evidence alone I assign to Lugalzaggisi the minimal date of 4000 B.C. My own personal conviction, however, is that he cannot have lived later than 4500 B.C.

3. That my determination of the age of Lugalzaggisi is not too high is proved by the discovery of an uninscribed vase of precisely the same material and characteristic shape[2] as most of the vases which bear Lugalzaggisi's inscription. It was found 1.54 m. below the pavement of Narâm-Sin, and must therefore considerably antedate the rule of the latter.

4. From palæographic and other reasons, I came to the conclusion above, that the inscriptions of Lugalzaggisi and of the other kings, patesis, etc., from Nippur grouped together with them, are surely older than Edingiranagin. Heuzey, on the basis of other arguments, had inferred that the stele of vultures and the reliefs of Ur-Ninâ are "surely older than Narâm-Sin." Hence it would follow, that if Heuzey's judgment of the age of these specimens of art is correct, also the monuments of Lugalzaggisi, etc., antedate Narâm-Sin. I am now in the position to prove the correctness of Heuzey's view beyond question. Since a specimen of the workmanship of the artists at Narâm-Sin's time was recently discovered (cf. Pl. XXII, No. 64), showing exactly the same high degree of execution as the script on his monuments, every Assyriologist is enabled to judge for himself as to the value of Heuzey's judgment. There are, however, a few fragments of a relief in clay lately discovered in Nippur, which must be regarded as the strongest evidence in favor of the French scholar's determination. While Heuzey declared Ur-Ninâ's and Edingiranagin's reliefs to be of greater anti-

---

[1] It is needless to quote passages from Mr. Heuzey's works in addition to those given on p. 257, note 6. In connection with his discussion of the age of the stele of vultures he makes the emphatic statement, "le type linéaire de l'écriture est assurément plus ancien que celui des inscriptions de Narâm-Sin, etc." (cf. *Les Origines Orientales*, p. 50).

[2] Haynes reported on this vase, August 10, 1895, expressing the hope that I might be able to use it in support of my theory as to the age of most of the other ancient vase fragments from Nippur. He found it covered with earth and black ashes. It consists of white calcite stalagmite and has a very characteristic shape never found at a later period in Nippur again. In general this class of vases resembles a flower-pot, the diameter at the top being larger than that at the bottom, while the walls frequently recede a little at the middle. The size of the above-mentioned vase is: h., 26.5; d. at the top, 18; at the bottom, 14.8; at the middle, 13.8 cm.

quity than Narâm-Sin's monuments, he characterized the relief which opens the splendid series of De Sarzec's finds (Pl. I, No. 1), and has several points of contact with the art exhibited in the stele of vultures, as "plus primitif, même que celui de la grossière tablette du roi Our-Nina" [De Sarzec, *l. c.*, Pl. 1, No. 2], and as "une œuvre d'une antiquité prodigieuse, un monument des plus précieux, que nous devons le placer avec respect tout à fait en tête des séries orientales, comme le plus ancien example connu de la sculpture chaldéenne." These words of a true master of his subject have found a splendid confirmation in the clay reliefs of Nippur just referred to, which in their whole conception and execution show a striking resemblance to the oldest specimen of art recovered from Tello. They were found 7–7.70 m. below the level of Narâm-Sin's pavement, and within about 1.50 m. of the lowest trace of Babylonian civilization.[1] Truly the genius and critical penetration of Heuzey could not have won a more brilliant victory.

5. In connection with my examination of the pre-Sargonic strata of Ekur, I twice called attention to the fact that baked bricks found below Narâm-Sin's pavement are plano-convex in form.[2] I might have added that no other form of baked brick has so far been discovered anywhere in the lowest strata of Nippur, and that these bricks as a rule bear a simple thumb mark upon their convex side. The form of these baked bricks, until the contrary has been proved, must therefore be regarded as a characteristic feature of all structures previous to the time of Sargon I and Narâm-Sin. It is quite in accordance with this view that the only inscribed bricks of Tello which show this peculiar form, bear the legend of Ur-Ninâ, whom on other evidence I placed before Sargon and Narâm-Sin.

6. We draw a final and conclusive argument from a door-socket of Sargon himself. In Part I, Pl. 14, Nos. 23–25, I published three brief legends of a king whom, influenced by Pinches's reading (Garde), I read Gande (pp. 28 ff.), and whom I regarded as identical with Gandash, the founder of the Cassite dynasty. All that I brought forward in favor of this identity I herewith withdraw; when I wrote those

---

[1] Cf. above, p. 240, note 2. They will be published in Series B of the expedition work edited by myself.

[2] The bricks of the ancient curb around the altar, p. 238, and the bricks of the ancient arch, p. 240. In his report of Oct. 26, 1895, Haynes refers to the discovery of a terra-cotta floor with a rim a little below the pavement of Narâm-Sin. He regards it as a combination of bath and closet, "proving that the present customs and methods of preparing the body for worship, as practiced by Moslems [in the immediate neighborhood of their mosques], is of very great antiquity. The drainage from this floor was conducted into a large vertical tile drain, which is 2 m. long and has an average diameter of 85 cm." This tile drain is "supported by a double course of bricks, plano-convex in form, with finger marks on the convex side." For a specimen of Ur-Ninâ's bricks cf. De Sarzec, *l. c.*, Pl. 31, No. 1. Specimens of this class of Nippur bricks were given by Peters in *The American Archæological Journal* X, p. 34 (two drawings from the hand of the late Mr. Mayer, † 20 Dec., 1894, in Bagdad). The peculiar shape of these bricks in the arch is scarcely distinguishable on Pl. XXVIII of the present work.

pages, I was still somewhat influenced by the current view of Assyriologists, that later kings occasionally imitated older patterns in their script. Since then I have completely shaken off this old theory as utterly untenable when contrasted with all the known facts of Babylonian palæography. The observation, however, which I made on p. 29, note 2, that the characters represent the peculiarities of Ur-Ninâ's inscriptions was entirely correct. Since then a large number of vase fragments have been excavated, by which I was enabled to confirm and strengthen my previous judgment based upon the study of a few squeezes of badly effaced inscriptions and to analyze the palæographic peculiarities of this whole class of ancient texts completely. I arrived at once at the result that the three legends published on Pl. 14 were written by Lugal-kigub-nidudu, "lord of Erech, king of Ur," who left us No. 86. Among other gifts, such as vases, dishes, etc.,[1] this sovereign presented a number of unhewn diorite, calcite, stalagmite and other blocks[2] to the temple as raw material for future use.[3] At the time of Bur-Sin II several of these blocks, of which one is published on Pl. XVII, were still unused.[4] They had been handed down from a hoary antiquity and scrupulously preserved for c. 1500–2000 years in the temple archive. Bur-Sin II selected a diorite block from among them, left the few words of its donor respectfully on its side,[5] turned it into a door-socket, wrote his own inscription on its polished surface and presented it in this new form to the temple. But something similar happened many hundred years before. According to Part I, p. 29, section 1,[6] the same rude inscription is scratched upon the back side of a door-socket of Sargon I. From the analogous case just treated it follows that Lugal-kigub-nidudu must have lived even before Sargon I, and consequently that all other inscriptions which have the same palæographic peculiarities as his own can only be classified as pre-Sargonic.

[1] Cf. Pl. XVIII, 40–48.

[2] Cf. Part I, p. 29.

[3] These blocks received therefore only a kind of registering mark scratched merely upon their surface (*Dingir En-lil(·la) Lugal-ki-gub-ni-dudu (ne) a-mu-na-shub*, "To Inlil L. presented (this" = *ne*)). The inscription on the block, Pl. XVII, No. 39, had originally 8 li. according to the traces left. On the diorite blocks these inscriptions are well preserved; on the calcite blocks however, whose surface corroded and crumbled in the course of six millenniums, they have suffered considerably. Cf. on the whole question of presenting stones as raw material to the temple, Hilprecht in *Z. A.* VIII, pp. 190 ff.

[4] As shown above.

[5] Cf. The curses on the statue B of Gudea, col. VII, 59 ff., on the door-sockets of Sargon, Pl. 1, 12 ff., Pl. 2, 13 ff., on the lapis lazuli block of Kadashman-Turgu, Pl. 24, pp. 14–20. In the latter case the lapis lazuli was likewise presented as raw material to be used in the interest of the temple. But the inscription—this was the intention of the donor—was to be preserved (a thin piece of lapis lazuli being cut off, cf. Pl. XI, No. 25) in remembrance of the gift.

[6] Cf. Part I, "Table of Contents," p. 47.

CONTENTS AND HISTORICAL RESULTS.

In the briefest possible way I will indicate the general results which I draw from a combined study of the most ancient Nippur and Tello inscriptions. With the very scanty material at my disposal this sketch can only be tentative in many points. For every statement, however, which I shall make, I have my decided reasons, which will be found in other places.[1]

At the earliest period of history which inscriptions reveal to us, Babylonia has a high civilization and is known under the name of *Kengi*, "land of the canals and reeds,"[2] which includes South and Middle Babylonia and possibly a part of the North. Its first ruler of whom we know is " *En-shagsag-ana*, lord of Kengi."[3] Whether he was of foreign origin or the shaykh of a smaller Babylonian " city " which extended its influence or the regular descendant of the royal family of one of the larger cities, cannot be decided. It is therefore impossible to say whether he belonged to the Sumerian or Semitic race, or traced his origin to both. That the Semites were already in the country results, aside from other considerations,[4] from the fact that the human figures on the stele of Ur-Enlil, which belongs to about the same period,[5] show the characteristic

---

[1] In *Assyriaca*, part II, in *Z. A.*, and in response to a repeated invitation from the President and Secretary of the Philosophical Society of Great Britian, in the *Transactions* of the latter society, where I expect to give a more complete sketch of the political and social conditions of ancient Babylonia.

[2] Cf. No. 90, 4 (also No. 87, col. II, 21) and above p. 252, note 9.

[3] His inscriptions (Nos. 90–92) have the oldest form of *mu*, have older forms for *sag* and show other characteristic features of high antiquity. His name signifies "lord is the king of heaven."

[4] Cf. for the present only the important argument drawn from Lugalzaggisi's inscription No. 87, col. III, 36. Here we have the same writing *DA-UR*, which from the inscriptions of Nebuchadrezzar II and other latest Babylonian kings, is known to be a Semiticism for *dâru*. Cf. Delitzsch, *Assyrisches Handwörterbuch*, p. 213.

[5] It has the most ancient forms for *dam* and *mu* and shows a very characteristic feature of the oldest period of writing by contracting the name of *Nin-din dug(-ga)*, or *Ba'u* (cf. above p. 252) into a monogram. The primitive style of art, and such details as the headdress of the god, the short garment of the two persons following the sheep and goat, the nakedness of Ur-Enlil, the fact that his figure and the other two have their hair shaved off, corroborate my determination of the age of this monument. On the other hand, this stele and No. 38 of the same plate, which doubtless belongs to the same age, show us a real Old Babylonian master, who produced a beautiful ensemble with a few simple lines, and knew how to breathe life into his very realistic but very graceful figures. Cf. the great skill he exhibits in his drawing of the graceful outlines of a gazel, and his remarkable knowledge of animal locomotion ! The two animals in No. 37 "represent very characteristically two species, the near one a goat and the far one a sheep. The goat shows more characteristics of the wild species of Eastern Persia and Afghanistan than of the Persian, and so may be a domestic hybrid between the two (*i. e.*, *Capra falconerii* and *Capra ægagrus*). The sheep is probably also derived from Eastern Persia and is perhaps the ' urial ' *Ovis vignei*, which is an ally of the domestic sheep. It has resemblance also to the Armenian wild sheep *Ovis gmelinii*, but the rugosity of the horns is too great, and the lines are too vertical " (communication from my colleague, Dr. Edward D. Cope, Professor of Zoölogy and Comparative Anatomy, who kindly examined the monument).

features of a mixed race.[1] The capital of this early kingdom is likewise unknown.[2] In all probability it was Erech.[3] The religious centre of Kengi was the sanctuary of Inlil at Nippur.[4] It stood under the especial care of every ruler who claimed supreme authority over the country, and who called himself *patesi gal Inlil*,[5] to define his position as being obtained by divine authority. The chief local administrator of the temple in Nippur seems to have had the title *damkar gal*.[6] This I infer from my analysis of the meaning of *damkar* and from the inscriptions of Nos. 94 and 95 in connection with No. 96, where a certain Aba-Inlil (= *Kishit-Bêl*) who has the title of *damkar*, presents a vase to Ninlil "for the life of Ur-Inlil, patesi of Nippur."[7] Ur[8] and Larsam[9] and doubtless other places whose names are not yet known from inscriptions, were prominent cities in this early Babylonian kingdom. They had their own sanctuaries, which stood under the control of a *patesi*. This title characterizes its bearer, according to his religious position, as sovereign lord of a temple and chief servant of the god worshiped in it. The fact that a patesi, in addition, often occupied a political position as king or governor, does not interfere with this view. He is first of all the highest official of his god, representing him in his dealings with his subjects; in other words,

[1] Prof. Cope wrote me on this subject : "The shortness of the jaws however is certainly not a Semitic character in human faces, and this character renders the physiognomy very peculiar. The hooked nose and large eyes on the contrary are Semitic. As a result I should say the figures represent an Aryan race with some Semitic tendencies. The identification of such a race is of much interest [indeed it is of vital importance for the whole Sumerian question ! —II.]. The people evidently have no Mongolian tendencies."

[2] It may have stood in No. 90, 5, *lugal* . . . . , which is only preserved in part. The traces do not point to the ideogram of *Unug*, more to *kalama*.

[3] Cf. Nos. 86, 4–14 ; also the fact that Erech is the capital of Lugal kigub-nidudu and Lugalzaggisi and is prominently mentioned in Edingiranagin's inscriptions. Cf. also Hommel, *Geschichte*, p. 206, and especially p. 300, observe the important position which Erech holds in the titles of the kings of the dynasty of Isin *en (shega) Unugaki* [N. B. Winckler's reading of Part I, No. 26, 3, as *Sin-ga-mil*, is an absolute palæographic impossibility. If anything, the reading of this line as *Unugki-ga-ge* is sure beyond question (against Winckler, *Altorientalische Forschungen* III, p. 274)].

[4] Cf. above, p. 236, and among other points, especially No. 87, col. I, 36–41.

[5] Cf. No. 87, col. I. A similar title occurs in the inscriptions of Tello, *patesi gal Ningirsu* (Entemena and his son Enanatuma). Apparently at an early time the god Ninib received the title *patesi gal Inlil* (Pl. 55, Obv. 17), and the kings and governors were satisfied with the title *patesi Inlil*.

[6] Cf. No. 94 : 1. *Dingir Nin-din-dug*, 2. *Ur-dingir En-lil*, 3. *dam-kar gal*, 4. *a-mu shub*, "To Ba'u Ur-Enlil the chief agent (*scil.* of Inlil) devoted (it)." The current translation of *damkar*, "merchant," is too narrow in many passages. Cf also No. 95 : 1. [*Dingir N*]*in-din-dug-ga* 2. *Ur-Ma-ma* 3. [*d*]*am-kar* 4. [*iluE*]*n-*[*lil*] 5. [*a-mu-na shub*], "To Ba'u Ur-Mama, agent of Enlil presented it." For *dingir Ma-ma* cf. the ideogram of Gula, *dingir Me-me* in later texts (*e. g.*, Strassmaier, *Cambyses*, 145, 3) and the goddess Mami II *R.* 51, 55a, and in old Babylonian contracts (the last two references I owe to Jensen). From the fragment of an inscribed stone in Bagdad I copied the phrase "*dam kar dingirDUN-GI*, preceded by the titles of a king of the second dynasty of Ur, and followed by *dingir Uruki-ka*.

[7] Cf No. 97, which seems to have been devoted by this very [Ur]-Enlil, patesi of Nippur, to Bêl.

[8] Cf. Nos. 86 and 87, col. II, 30–32, mentioned also by Edingiranagin.

[9] Cf. No. 87, col. II, 33–37.

he is the legitimate possessor of all the privileges connected with this title. These privileges vary according to the sphere of power which a god exercises beyond the limits of his temple or city, and depend chiefly upon the popularity of his cult, the personal devotion and energy of his human representative, and, more than anything else, upon the strength and valor of the city's army. In order to define them accurately, it is first of all necessary to determine the political power of the god's city in each individual case. As soon as we have a clear conception of the latter, we have the key to a correct understanding of the position and privileges of its patesi. But the title itself does not express any reference either to the political dependence or independence of its bearer.[1]

A troublesome enemy of Babylonia at this early period was the city of Kish, which therefore did not form part (any longer?) of Kengi proper. It had apparently its own peculiar cult and stood under the administration of a patesi,[2] who was eager to extend his influence far beyond the limits of his city, and sought every opportunity to encroach upon the territory of his southern neighbor. For Kish is styled *ǵul shaǵ*[3] " wicked of heart," or *ga ǵul*[4] " teeming with wickedness." The very fact that one

---

[1] Winckler, *Altorientalische Forschungen* III, pp. 232ff. gives a very good analysis of the relation of a god to his city and of the origin and growth of Oriental states in general, and of the Babylonian kingdom in particular, but his view as to the meaning and use of the word *patesi* is entirely incorrect ("die gebräuchliche Bezeichnung für die unterworfenen Könige ist in Babylonien *patesi*," p. 234). An interesting monument from Tello, recently published by Heuzey in *Revue d'Assyriologie*, serves as an excellent illustration of the correctness of my definition, which I share with Tiele (*Z. A.* VII, p. 373), Hommel (*Geschichte*, p. 294 f.) and other Assyriologists. The inscription to which I refer had defied the united efforts of Oppert, Heuzey and myself for a long while. But I am now able to offer the following correct interpretation. *Sa! Lugal Kish, sanga ᵈⁱᵘNin-su-gir (sic!) ᵈⁱᵘNin-su-gir mu-gin, Lugal-kurum-zigum pa-te-si Shir-[pur]-l[aki]*, " Decision ! Ninsugir has appointed the king of Kish as priest of Ninsugir. Lugal-kurum zigum is patesi of Shirpurla." This valuable document is important in more than one way. The whole phraseology seems to be Semitic rather than Sumerian (cf. also *sanga* artificial ideogram composed of *sa + ga*). The name means *Sharru-kurumat-shamê*, " The king is food of heaven " (" Der König ist Himmelsspeise "). A foreign conqueror of Shirpurla, who is already a king, in addition styles himself patesi of Lagash, expressly declaring that Ningirsu himself, the highest god of the city, called him to fill this office. The condition of affairs is here plain. The conqueror seeks to represent to the people and to the priesthood his violent act as having been committed in the service of their god and carrying out his decision. Therefore he does not call himself king—which he already was—nor *patesi* in the sense of our governor, because he cannot designate himself as his own subject, but *patesi* as the highest official of the god Ningirsu, in the care of his temple and in the administration of that territory over which Ningirsu ruled ; in other words, as the legitimate possessor of all the privileges which, up to the time of his conquest, had been connected with this title. Cf. Hilprecht, *Recent Research in Bible Lands*, pp. 71 ff.

[2] Cf. Nos. 108 and 109 (portions of the same vase). The beginning (No. 108) is to be restored as follows : 1. ᴰⁱⁿᵍⁱʳZa-[ma-ma] 2. U-dug- . . . . 3. pa t[e-si] 4. Ki[shki].

[3] No. 92, 4.

[4] No. 102, 4. *Ga* is written phonetically for *ga(n)*, Brünnow, *List* 4039, as becomes clear from a comparison of No. 113, 4 with 8 and No. 112, 4. No. 112 reads as follows : 1. ᴰⁱⁿᵍⁱʳNin-lil 2. ᴰⁱⁿᵍⁱʳEn-lil-la(l) 3. dumu ad-da-ge 4. ga-til-la-shu 5. nam-ti 6. dam-dumu-na-shu 7. a-mu-na-shub, " To Ninlil and Inlil the son of the ada (scil. of the temple of Inlil, No. 113, 6f.) presented it for abundance of life, for the life of his wife and child." Apparently a son

patesi of Kish presented a large sandstone vase to Inlil of Nippur, shows us that temporarily he was even in possession of an important part of Kengi, including the sanctuary of Bêl. Enshagsagana himself waged war against his northern enemy, and presented the spoil of this expedition to Inlil of Nippur.[1] The same was done by another king of Kengi, who lived shortly before or after. He infested Kish and defeated or even captured its king, Enne-Ugun.[2] " His statue, his shining silver, the utensils, his property," he carried home victoriously, and deposited in the same sanctuary as his

was born unto him, and the happy father presented a vase to the temple. Cf. Jensen in Schrader's *K. B.* III, part 1, p. 25, II (where Jensen and Amiaud, however, misread the name of the donor. As the separating lines clearly prove, the name is not *Ur-Enlil* but *Ur-Enlil-dabi-dudu*). No. 113 reads : 1. $^{Dingir}Nin$-*lil-ra* 2. *Uru-na-bada-bi* 3. *sang* (Amiaud et Méchineau, *Tableau,* No. 134) $^{dingir}En$-*lil* 4. *gan-til-la-shu* 5. *Ur-Simug* (Amiaud et Méchineau, *l. c.,* No. 117) *-ga* ($^{dingir}Simuga=$ Ea !) 6. *dub-sar ada* 7. *e* $^{dingir}En$-*lil-ka-ge* 8. *ga-ti-la-shu* 9. *nam-ti* 10. *ama dug*(sic!)-*zi-shu* 11. *nam-ti* 12. *dam-dumu-na-shu* 13. *a-mu-na-shub,* "To Ninlil Urunabadabi, priest of Inlil, for abundance of life, and Ur-Simuga ('servant of Ea'), scribe of the ada of the temple of Inlil (*ada e* identical with the frequent title of the later contract literature *abu biti*!), for abundance of life presented it for the life of his (distributive = their !) good and faithful mother, and for the life of his (their) wife and child." Apparently two brothers who held two different positions in the temple of Bêl presented together this beautiful vase for their mother, wives and children. Cf. also No. 106: 1. $^{Dingir}Nin$-*d*[*in*]-*dug-ga* 2. *Nin-en-nu* (cf. *Lugal-en-nu,* No. 114, 5) 3. *ga-til-la-shu* 4. *a-mu-na*[*-shub*], "To Ba'u *Ninennu*(for *en-nun = naṣaru*!) presented it for abundance of life." My constant transliteration of the postposition "*ku*" by *shu* needs a word of explanation. I believe with Jensen, that no Sumerian postposition *ku* exists, and that the old Babylonian sign of this postposition transliterated by *ku* is rather identical with the character in Part I, Pl. 1, 13 ; Pl. 2, 13, which I identified as *shu* (*l. c.,* pp. 13 f.).

[1] Cf. Nos. 91 and 92, which supplement each other : 1. [$^{Dingir}E$]*n-lil-la* 2. *En-shag-sag-an-na* 3. *nig-ga Kish*ki 4. *ḡul shag* 5. *a-mu-na-shub,* "To Inlil E. presented the property of Kîsh, wicked of heart (referring to Kîsh)." In connection with this text I call attention to the fact that the word *namrag* "spoil," the etymology of which was obscure (cf. Part I, p. 21) is purely Sumerian, being composed of *nam+ri+ag* (V R. 20, 13c), corresponding to Assyrian *shallatu shalalu* (cf. Delitzsch, *Assyr. Gram.,* §§ 73, 132), a synonym of *shallatu* "spoil."

[2] Several vase fragments mention this event, but the whole inscription cannot yet be restored from them. Nos. 103 + 110 belong to the same vase. Nos. 104 and 105, which contain portions of the same inscription and supplement part of the text, belong to two other vases. The fragment of a fourth vase, No. 102, contains part of the same inscription. For C. B. M. 9297, which has remnants of l. 1–4 of No. 102, agrees in thickness, material and characters of writing entirely with Nos. 103 + 110 and belonged doubtless to the same vase. No. 105 had a briefer inscription than the rest. Of the longer inscription the beginning is wanting, the first preserved portion, No. 103, is to be supplemented by No. 104, to be continued by No. 102, 2, and (after a break of several lines) to be closed with No. 110. I restore the inscription as follows : 1. [$^{Dingir}En$-*lil-la* 2. [*lugal kur-kur-ra* 3. Name of the king 4. [*en Ki-en-gi*] 5. (No. 103 begins) [*lu*]*gal . . . .* 6. *ud* ^{dingir}[*En-lil-li*] 6. *ma-na-ni-ḡun-a* (cf. No. 86, 1–5) 7. *Kish*ki 8. *mu-ḡul* 9. *En-ne-Ugun* (Brünnow, *List* 8862, cf. Jensen in *Z. A.* I, p. 57f.) 10. *lugal Kish*ki 11. *mu-dur* 12. *lugal erim* gish*BAN*ki-*ka-ge* 13. *lugal Kish*ki-*ge* 14. *uru-na ga* (written phonetically $=$ *gan,* Brünnow, *List* 4039, for cf. No. 113, 4, with 8 and No. 112, 4) *ḡul* 15. *nig-ga* 16. *. . . . bil* 17–18 (or more) wanting 19. *mu-ne-gi* 20. *alana-bi* (observe the peculiar sign for *bi* in Nos. 105 and 110 !), 21. *azag-zagina-bi* 22. gish *nig-ga-bi* 23. $^{dingir}En$-*lil-la* 24. [*E*]*n-lil*ki-*shu* 25. *a-mu-na-shub* ["To Inlil, lord of lands, N. N., lord of Shumer (king of Erech)]—when he had looked favorably upon him ($=$ *nashû sha êni,* Brünnow, *List* 10545), he infested Kîsh, he cast down (or bound? cf. Jensen in Schrader's *K. B.* III, part 1, p. 48) Enne-Ugun, king of Kîsh ; the king of the hordes of gish*BAN*ki, king of Kîsh—his city teeming with malignity, the property . . . . he burned, ... he brought back, and his statue, his shining silver, the utensils (*isu =* ânu, II R. 23, 9 c.f.), his property, he presented unto Inlil of Nippur." The reading of the name of the king of Kîsh is of course only provisional. He was apparently a Semite.

predecessor. It is highly interesting to learn from the votive inscription with which the Babylonian ruler accompanied his gift (No. 102), that the king of Kish apparently had connections with the city of ^{gish}BAN^{ki}. For he is styled "king of the hosts of ^{gish}BAN^{ki}, king of Kish." In other words, we find the two mentioned cities in exactly the same close association as they appear on Edingiranagin's famous stele of vultures. It is therefore evident that the king of Kish was not only an ally of ^{gish}BAN^{ki}, but as commander of an army of this country, was in all probability himself a native of ^{gish}BAN^{ki}. In other words, I infer from this and other passages, that Kish (which I believe formed originally part of Kengi) at this early time was already under the control of a foreign people, which came from the North, appeared at the threshold of the ancient Sumerian kingdom of *Kengi*, and was constantly pushing southward. Kish formed the basis of its military operations, and at this time was, in fact, the extreme outpost of the advancing hordes of ^{gish}BAN^{ki}, serving as a border fortification against Kengi. The success of the Babylonian monarch who defeated Enne-Ugun, cannot have lasted very long. For another king of Kish, Ur-Shulpauddu,[1] presented several inscribed vases "to Inlil, lord of lands, and to Ninlil, mistress of heaven and earth, consort of Inlil" (No. 93), and was therefore in the possession of Nippur. He must have dealt a fatal blow to the kingdom of Kengi, for besides his usual title *lugal Kish* he assumed another, which unfortunately is broken away.[2] To judge from the analogy of other inscriptions of this period, I have no doubt it contained the acquired land or province of which Kish had now become the capital,[3] scarcely, however, *Kengi* itself. How long he ruled, how far his kingdom extended, and whether he was able to hold his conquests, we do not know. So much is certain, the great centre in the North which controlled the movements of its warriors in the South, continued to send out its marauding expeditions against Babylonia. And even if a temporary reaction occasionally should have set in, the weakened South could not withstand the youthful strength and valor of its northern enemies for any length of time. At last ^{gish}BAN^{ki} was prepared to deal the final blow to the ancient kingdom of Kengi, however little of it there may have been left. The son of "Ukush, patesi of ^{gish}BAN^{ki},[4] was this time himself the chief commander of the approaching army. Erech opened its doors, and the rest of Babylonia down to the Persian gulf fell an easy prey to the conquering hero. A hero indeed, Lugalzaggisi was, if we can trust his own long inscription

---

[1] "Servant of Shulpauddu." The same name occurs occasionally in the early contracts of Nippur and Tello. Cf. Scheil in *Receuil* XVII, p. 41.

[2] Traces of *lugal* are clearly visible in l. 8.

[3] No. 87, col. I, 5.

[4] *I. e.,* "The king is filled with unchangeable power." Cf. *Nimrod Ep.*, 12, 39 ; *Gilgamesh yitmalu enůku.* The name is possibly to be read Semitic.

of 132 lines,[1] carved over 100 times on as many large vases, which he presented to the old national sanctuary of the country in Nippur.

The titles themselves with which he opens his dedication are a reflex of the great achievements he could boast of: Col. I, 3. " Lugalzaggisi, 4. king of Erech, 5. king of the world, 6. priest of Ana, 7. hero 8. of Nidaba, 9. son of Ukush, 10. patesi of *gish*BAN[ki], 11. hero 12. of Nidaba, 13–14. he who was favorably looked upon by the faithful eye of Lugalkurkura (*i. e.*, Inlil), 15. great patesi 16. of Inlil, 17. unto whom intelligence was given 18. by Enki[2] (= Ea), 19. he who was called (chosen) 20. by Utu, 21. sublime minister[3] 22. of Enzu (= Sin), 23. he who was invested with power 24. by Utu,[4] 25. fosterer of Ninna, 26. a son begotten 27. by Nidaba, 28. he who was nourished with the milk of life 29. of Nin-ḫarsag,[5] 30. servant of Umu, priestess of Erech, 31. a slave brought up 32. by Nin-a-gid-ga[6]-du, 33. mistress of Erech, 34. the great *abarakku* of the gods."[7] He was one of the greatest monarchs of the ancient

---

[1] It is the longest complete inscription of the fourth and fifth pre-Christian millenniums so far obtained from Babylonia, and as a historical document of this ancient period it is of fundamental importance. The text published on Pls. 38–42, No. 87, was restored by myself from 88 fragments of 64 different vases under the most trying circumstances. The work was just as much a mathematical task as it was a palæographical and philological problem. On the basis of palæographical evidence I selected c. 150 pieces out of a heap of c. 600 fragments and particles. Then I succeeded in placing the five fragments on Pl. XIX, No. 49, together. By doing this I obtained the beginnings and ends of each column. I noticed that the lines of each of the first two columns must be identical, as the separating lines run from the first to the last column. The difference of the numbers of lines between the second and third lines I could easily determine by a simple calculation. It was more difficult to find out the exact number of lines of which the first and second columns originally consisted. By calculating the original circumference, and making a number of logical combinations, I arrived at the conclusion, which finally proved to be correct, that each of the first two columns had forty-six and the third only forty lines. Then followed the tedious work of arranging the little fragments and determining their exact position, although often enough not more than a few traces of the original characters were left to guide me. I had the complete translation prepared for this volume, but I am obliged to withdraw it from want of space. In the previous and following pages nearly two-thirds of the whole inscription have been treated, according to the passages needed. A complete coherent transliteration and translation will be found in another place very soon. Since the restoration of my text, Haynes has found many duplicates, which in every case confirmed the correctness of my arrangement. Col. III, 25f. can now be restored completely.

[2] Cf. Jensen in Schrader's *K. B.* III, Part 1. The titles of Lugalzaggisi are not unsimilar to those of kings and patesis of Tello.

[3] Cf. above, p. 255, note 6.

[4] One expects rather the ideogram for *shakkanakku* (Brünnow, *List* 9195). *Ne* ("power") + *gish* ("man") apparently is its synonym. Cf. *sag-gish*, I R., 2, No. 5, 1 (and 2), 3 ; the present work, Part I, No. 81, 7.

[5] Literally "ate" (*akâlu*) or "was filled with " (*shuznunu*).

[6] The variant is a peculiar form of *ga* (not = *igi*), cf. col. III, 21, 23 and variants.

[7] No. 87, col. I, 1. *Dingir*En-lil 2. *lugal kur kur-ra* 3. *Lugal-zag-gi si* 4. *lugal Unugki-ga* 5. *lugal kalam-ma* 6. *shib An-na* 7. *galu maǵ* 8. *dingirNidaba* 9. *dumu U-kush* 10. [*pa-t*]*e-si gishBANki* 11. *galu maǵ* 12. *dingirNidaba-ka* 13. *igi zi da* 20. *dingirUtu* 21. *luǵ maǵ* 22. *dingirEn-zu* 23. *ne-gish* 24. *dingirUtu* 25. *ú-a dingirNinna* 26. *dumu tu-da* 27. *dingirNi-daba* 28. *ga zi ku-a* 29. *dingirNin-ḫar sag* 30. *galu dingirUmu sanga Unugki-ga* 31. *sag eǵi-a* 32. *dingirNin-a-gid-ǵu-du* 33. *nin Unugki-ga-ka* 34. *iti* (?) *maǵ* 35. *dingir-ri-ne-ra.*
bar-ra 14. *dingirLugal-kur-kur-ra* 15. *pa te-si gal* 16. *dingirEn-lil* 17. *gish-PI-SHU-sum-ma* 18. *dingirEN-KI* 19. *mu-pad-*

East, and yet his very name had been forgotten by later generations. He lived long before Sargon I founded his famous empire, and he called a kingdom his own which in no way was inferior to that of his well-known successor, extending from the Persian Gulf to the shores of the Mediterranean. I quote the king's own poetical language: "When Inlil, lord of the lands, invested Lugalzaggisi with the kingdom of the world and granted him success before the world, when he filled the lands with his renown (power) (and) subdued (the country) from the rise of the sun to the setting of the sun—at that time he straightened his path from the lower sea of the Tigris and Euphrates to the upper sea and granted him the dominion of everything (?) from the rise of the sun to the setting of the sun and caused the countries to rest (dwell) in peace." [1] It becomes evident from this passage, in which Lugalzaggisi declares himself to have been invested with the kingdom of the world by Inlil of Nippur, "lord of the lands," that only Nippur can have been the ancient seat of the *sharrût kibrat arba'im*, which manifestly is but the later Semitic rendering of the ancient Sumerian *nam-lugal kalama*. I have examined all the passages in the fresh light of this text and find that Nippur fulfills by far better the required conditions than Kutha or any other city which has been proposed in Northern Babylonia. But, be it remembered, to the early kings of Babylonia this title meant more than a mere possession of the city whose god claimed the right of granting the *sharrût kibrat arba'im*. Down to the time of Hammurabi only those[2] laid claim to this significant title who really owned territory far beyond the north and south of Babylonia, who, in the Babylonian sense of the word, had conquered a *quasi* worldwide dominion, defined by the four natural boundaries (Part I, p. 25). The later Babylonian and Assyrian inscriptions are of value for the determination of the meaning of this title at their own time, but they have little importance for the question as to its origin and earliest localization, if the title must be localized at all hazards.

According to the manner of usurpers,[3] Lugalzaggisi retained Erech, the old metropolis of the country, as his own new capital of this first great Oriental state, of which Kengi became now the chief province. Babylonia, as a whole,[4] had no fault

---

[1] Col. I, 36. *Ud* dingir*En-lil* 37. *lugal kur-kur-ra* 38. *Lugal-zag-gi-si* 39. *nam-lugal* 40. *kalam-mà* 41. *mà-na-sum-ma-a* 42. *igi kalam-ma-ge* 43. *si mà-na-di-a* 44. *kur-kur(a)ne na* 45. *mà-ni-sig ga-a* 46. *Utu e(a)-ta.* Col. II, 1. *Utu shu(a)-shu* 2. *gu mà-na-gar-ra-a* 3. *uda-ba* 4. *a-ab-ba* 5. *sig-ta-ta* 6. *Idigna* 7. *Buranunu*(without determ.)-*bi*(= "and") 8. *a-ab-ba* 9. *igi nim-ma-shu* 10. *gira-bi* 11. *si-mi-na-di* 12. *Utu e(a)-ta* 13. *Utu shu(a)-shu* 14. [dingir*E*]*n-lil li* 15. . . . . *nin* 16. . . . . . *mu-ni-dug* 17. *kur kur(a) ú sal-la* 18. *mu-da-na.*

[2] Of Dungi we know too little to call him an exception. Of the kings of the second dynasty of Ur, who assumed the proud title, we know now from Pls. 55 and 58 (cf. above, p. 246 and note 4) that they had made conquests as far as Syria and Elam.

[3] Well stated by Winckler, *Altorientalische Forschungen* III, p. 234.

[4] Cf. col. II, 19. *kalam-mà* 20. *a-ᶜul-la mu-da-gà* (= *shakânu*) 21. *bar-bar Ki-en-gi* 22. *pa-te-si kur kur-ra*, etc., etc.

to find with this new and powerful régime. The Sumerian civilization was directed into new channels and prevented from stagnation; the ancient cults between the lower Tigris and Euphrates began to revive and its temples to shine in new splendor. Erech, Ur,[1] Larsa[2] and Nippur[3] received equal attention from their devoted patesi. But first of all, ^{gish}BAN^{ki} itself, the native city of the great conqueror, was raised by his energy and glory to a position of unheard-of influence and political power. Lugalzaggisi stands out from the dawn of Babylonian history as a giant who deserves our full admiration for the work he accomplished. He did not appear unexpectedly on the scene of his activity. We had been prepared for the collapse of the ancient monarchy on the Persian Gulf, with its long but unknown history, by the preceding invasions and victories of the Northern hordes to which he belonged. And yet when suddenly this great empire of Lugalzaggisi stands before our eyes as a *fait accompli*, we can scarcely conceive, whence it came and how it arose.

There is no doubt in my mind that Lugalzaggisi's achievements in Babylonia represent the first signal success of the invading Semites from the North. On the previous pages we have seen how these hordes were pushing gradually southward. After for a number of years they had concentrated their attacks upon the border fortifications of Northern Babylonia and had established a military station and kingdom in Kish, it was but a question of time when the whole country in the South had to succumb to their power. The oldest written monuments of Babylonia do not designate these enemies by any single definite name: they are the hordes of the city of ^{gish}BAN^{ki} and Kish combined, apparently but two centres of the same powerful people which was roaming over the fertile steppes of Mesopotamia, and whose chief stronghold doubtless was ^{gish}BAN^{ki}. What ancient city, then, is this ^{gish}BAN^{ki}? That we have not to place it "in Susian territory," as Maspero[4] is tempted to do, is beyond question. The ideogram for *lugal* on an inscribed object of Tello and presented by a king of ^{gish}BAN^{ki} (De Sarzec, *l. c.*, Pl. 5, No. 3), points with necessity to the north for the location of our city. As this peculiar form of the character for *lugal* so far has only been found in such cuneiform inscriptions as contain Semitic words written phonetically, or in other texts which are written ideographically, but, on the basis of strong arguments[5] must be read as Semitic, we are forced to the conclusion that this charac-

---

[1] Col. II, 30–32. *Urum^{ki}-mà guda-gim sag-ana-shu mu-um-gur*, "Ur like a steer he raised to the top of heaven."

[2] Col. II, 33–37. *Larsam^{ki} ur ki-ag ^{dingir}Utu-ge a-ne-ĝul-la mu-da-gà*. For ^{gish}BAN^{ki} cf. *ibidem*, 38–42.

[3] As becomes evident from his titles and from the extraordinary number of vases presented to Inlil.

[4] *The Dawn of Civilization*, p. 608. Cf. also Heuzey in De Sarzec, *l. c.*, p. 182.

[5] Cf. for the present above, p. 263, note 1. More on this subject and on "the Semitic influence in early cuneiform writing in general in another place. My above statement is the result of a complete and exhaustive examination of all the published cuneiform material in which the peculiar form of *lugal* occurs.

ter, while doubtless derived from the well-known Sumerian form, was invented and employed by a Semitic nation. Furthermore, I call attention to the important fact that Lugalzaggisi, who was surely a Semite,[1] shows his nationality in various ways, such as the use of certain phrases, which look very suspicious in an ancient Sumerian inscription,[2] and especially in his use of the ideogram da-ur, doubtless of Semitic origin (= dârû), for " eternal."[3] There is only one ancient place in Northern Mesopotamia which could have been rendered as "the city of the bow" ideographically by the Sumerians, namely **Harran**, with which ^{gish}BAN^{ki} is doubtless identical. For according to Arabic writers, especially Albîrûni (ed. Sachau, p. 204),[4] the ground-plot of Ḥarrân resembled that of the moon (i. e., the crescent or half-moon), and Sachau, who gave us the first accurate sketch of this city, finds it very natural that " Arabic writers could conceive the idea of comparing it with the form of the half-moon."[5] Excellent, however, as this Arabic description is, and valuable as it proves for our final location of ^{gish}BAN^{ki}, the ancient Babylonian ideographic rendering as " city of the bow " was a more faithful description of the peculiar way in which Ḥarrân was built than any other, as everybody can easily convince himself by throwing a glance upon Sachau's plan in his *Reise in Syrien und Mesopotamien*. This correct solution of a vexed problem becomes of fundamental importance for our whole conception of the history of the ancient East. First of all, I have furnished a better basis for Winckler's ingenious theory of the original seat of the *sharrât kishshati*. All that could be gathered from later historical sources, beginning with the end of the second millennium before Christ, Winckler brought together to formulate a view which never found much favor with Assyriologists and historians.[6] I opposed it myself[7] on the ground that his reasons proved nothing for the ancient time, because Ḥarrân was never mentioned in a text before the period just stated, and that in view of the total absence of a single

[1] If he did not adopt a Sumerian name when ascending the throne of Kengi and of the "kingdom of the world," which is very probable, the name of the king must be read something like *Sharru-mâli-emûḳi-kênu* (emûḳu is masc. and fem. in the singular). But the name cannot be regarded as the prototype of Sargon I (= *Sharru-kênu*), because, aside from other reasons, this kind of abbreviation of a fuller name is without parallel in the history of Assyrian proper names. They are abbreviated at the beginning or end, but not in the middle. Cassite names, etc., are foreign names.

[2] Cf., e. g., "from the lower sea of the Tigris and Euphrates to the upper sea," "from the rising of the sun to the setting of the sun" and others, which remind us forcibly of the phraseology of the latest Assyrian monarchs.

[3] Col. III, 36. *da-ur ̮ge-me*, "he may pronounce (speak) forever!"

[4] Cf. also Mez, *Geschichte der Stadt Ḥarrân in Mesopotamien*, p. 9. The remark of the Arabic writer is therefore more than a "Treppenwitz," and is of great historical importance, showing us that not only the ancient Babylonians but other peoples were struck by the remarkable form in which Ḥarrân was built.

[5] Sachau, *Reise in Syrien und Mesopotamien*, p. 223.

[6] Cf. especially Winckler, *Altorientalische Forschungen* I, pp. 75ff.; III, pp. 201 ff.

[7] Part I, pp. 23 f. I was supported in this, e. g., by Jensen in *Z. A.* VIII, pp. 228 ff.

reference to this city in our whole ancient literature previous to 1500 B. C., we could not speak of it as the seat of a kingdom until we first proved that the city really existed. From the fact that (1) *Kish* and *Kish* (*shatu*) did not only sound alike but were even used interchangeably in the inscriptions,[1] (2) that many other ancient Babylonian cities (cf. Shirpurla)[2] are frequently written without a determinative, (3) that the city of Kîsh played a very important *rôle* in the inscriptions of Edingiranagin,[3] (4) that all the ancient empires arose from city kingdoms, and from several other considerations,[4] I inferred that *shar KISH* meant originally " king of Kîsh," a combination which Winckler himself regarded "naheliegend."[5] But notwithstanding the great importance which must be attached to the kingdom of Kîsh in connection with the final overthrow of the ancient empire of *Kengi*, Kîsh was not the principal leader in this whole conquest, but was controlled by a greater power in the North, Ḥarrân, as I have shown above. Having therefore demonstrated the existence of the city of Ḥarrân at the threshold of the fifth and fourth pre-Christian millenniums, which Winckler failed to do, although Edingiranagin's inscriptions, which necessarily formed the starting point of my operations, had been at his disposal for some time, and having furthermore indicated the powerful position which Ḥarrân must have occupied as the great Semitic centre of the ancient Orient, I am now prepared to accept Winckler's theory of the original seat of the *sharrût kishshati* without reserve. I regard the title as the Assyrian equivalent of the Sumerian *nam-lugal kalama*. In view of the leading part that Ḥarrân had taken in the establishment of the first " kingdom of the world " under Lugalzaggisi, Ḥarrân became the seat of the Semitic *sharrût kishshati* just as Nippur was the centre of the Sumerian *nam-lugal kalama*. When after many vicissitudes under Sargon I and Narâm-Sin finally the northern half of ancient Kengi, including Nippur, was definitely occupied by a Semitic population, which spoke and wrote its own language, the old Sumerian title *nam-lugal kalama*, which carried the same meaning for the inhabitants of Babylonia as *sharrût kishshati* did for

[1] Cf. Winckler, *l. c.*, pp. 144 f.

[2] In the inscriptions of Ur-Ninâ written without *ki*.

[3] Not only in his stele of vultures, but also in the inscription unearthed in London (*Proc. Soc. Bibl. Arch.*, Nov., 1890). Hommel was of the opinion (*Die Identität der ältesten babylonischen und ägyptischen Göttergenealogie*, p. 242), that the passage in the latter text escaped my attention. I simply had no use for it : (1) *lugal Kish an ki* is something entirely different from *lugal an-ub-da tab-tab-ba* or *lugal KISH*; for if it was possible to say so in Sumerian, it could only mean " king of the whole heaven and earth," which the king of course did not want to say. (2) The text does not offer this at all, but must be translated *lugal Kishki-bi-na-dib-bi,* "**and** the king of Kîsh," in other words *bi* is copula = "and," connecting *Kishki* with what stood before. Cf. in the present work, Pl. 87, col. II, 7 ("and " the Euphrates).

[4] Cf. Part I, pp. 23 f.

[5] *Altorientalische Forschungen* II, p. 145, note 1.

the Semites of Northern Mesopotamia, disappeared and was translated into the Semitic *sharrât kibrat arba'im*. The later Sumerian *nam-lugal* ^{an}*ub-da-tab-tab-ba* is nothing but a translation from the Semitic title back into the sacred Sumerian language by Semitic scribes of the third millennium B. C.

Not long after Lugalzaggisi's death a reaction seems to have set in. Sugir generally transliterated as Girsu, which Urukagina or one of his predecessors raised from the obscurity of a provincial town to the leading position in the new kingdom of Shirpurla, must be regarded as the centre of a national Sumerian movement against the Semitic invaders. "The lord of Sugir," *Nin-Sugir*, became the principal god, and his emblem – the lion-headed eagle with outspread wings, occasionally appearing in connection with two lions, which are victoriously clutched in its powerful talons[1]—became the coat-of-arms of the city and characterizes best the spirit of independence which was fostered in its sanctuary. Urukagina's successors, especially Ur-Ninâ, devoted their time to building temples and fortifying the city of Shirpurla and, as faithful patesis, impressed the power and glory of their warlike deity upon their subjects. The cult of Nin-Sugir cannot be separated from the national uprising which started from his sanctuary. Edingiranagin at last felt strong enough to shake off the obnoxious yoke of the Semitic oppressors of Kîsh and Harrân. The decisive battle which was fought must have been very bloody. The Sumerians won it, and they celebrated their victory, which restored a temporary power and influence over the greater part of Kengi to them, in the famous stele of vultures set up by Edingiranagin. Erech and Ur played a prominent part in this national war. The former retained its place as the capital of the *nam-en* (of Kengi), but Ur seems to have furnished the new dynasty, as I infer from No. 86.

Although No. 86 of my published texts belongs doubtless to the same general period as No. 87, a detailed examination of its palæographic peculiarities leads me to place it somewhat later, and to regard it as about contemporary with the inscriptions of the kings of Shirpurla, especially with those of Edingiranagin. We learn from it the following:[2] "When Inlil, the lord of the lands, announced life unto Lugal-kigubnidudu, when he added lordship to kingdom, establishing Erech as (the seat of) the lordship (the empire) and Ur as (the seat of) the kingdom, Lugal-kigub-nidudu presented this for the great and joyful lot (which he received) unto Inlil, his beloved

---

[1] Cf. Heuzey's treatise *Les Armoiries Chaldéennes*.

[2] Five different legends have been found of this ruler: (1) A brief legend of three lines (cf. Pl. 14), (2) one of seven or eight lines (cf. Pl. XVII No. 39), (3) one of nineteen lines, (4) an even larger one of c. thirty lines, (5) No. 88. Of the third class a fragment was excavated after the preparation of my plates, which contained the closing lines 17-19. The precise connection between the upper and lower portions on Pl. 37 cannot be given at present.

lord for his life."[1]   In Lugal-kigub-nidudu[2] and his son (?) Lugal-kisal-si[3] we have therefore the first representatives of the first dynasty of Ur. Ur-Gur and Dungi, etc., who lived about 1000 years later, must hereafter be reckoned as members of the second dynasty of Ur.[4] The relation of this dynasty to Edingiranagin is shrouded in absolute mystery. It is not impossible that its members ruled before him and were Semites who overthrew the dynasty of Lugalzaggisi.

How long the restored Sumerian influence lasted we do not know. Apparently the Semites were soon again in possession of the whole country. The old name *Kengi* continued to live as an ideogram in the titles of kings, but the name of Shumer, by which Southern Babylonia was known to the later Semitic populations, was derived from the city of *Sugir* or *Sungir*,[5] which was the centre of the national uprising of the South against the foreign invaders from Kish and Harrân. Sargon I finally restored what had been lost against Edingiranagin. In his person and work we see but a repetition of that which had happened under Lugalzaggisi centuries before. From the city of Agade,[6] which became the capital of the Sargonic empire, I derive Akkad, the name of Northern Babylonia. The names of Shumer and Akkad are therefore but the historical reflex of the final struggle between the Sumerian and Semitic races, and they were derived from the two cities which took the leading part in it.[7]

---

[1] 1. *Dingir En-lil.* 2. *lugal kur-kur(a)-ge.* 3. *Lugal-ki-gub-ni-du-du-ra*   4. *ud dingir En-lil-li* 5. *gu-zi mà-na-de a* 6. *nam-en* 7. *nam-lugal(a)-da* 8. *mà-na-da-tab-ba-a* 9. *Unugki-ga* 10. *nam-en* 11. *mu-ag-ge* 12. *Urumki-ma* 13. *nam-lugal* 14. *mu-ag-ge* 15. *Lugal-ki-gub-ni-du-du ne* 16. *nam gal-yul-la-da* 17. *dingir En-lil lugal ki-a[ga-ni* 18. *nam-ti-la-ni-shu* 19. *a-mu-na-shub*]. The use of *da* = *shu*, "unto, for," in this text is interesting, cf. l. 7 and l. 16. We meet the same use in No. 111 : 1. *Dingir Nin-din-dug-ga* 2. *ama nin* 3. *dam* 4. ff. . . . . 3 f. e. *Lugal-shir-ge* 2. f.c. *nam-ti* 1 f. c. *dam- dumu-na-* **da** *a-mu-shub.*

[4] "The king finished the place " = *Sharru-manzazu-ushaklil.*

[3] Or *Lugal-si-kisal, i. e.,* "The king is the builder of the terrace," *Sharru shâpik-kisalli.* From the close connection in which *Lugal-kigub-nidudu,* who left many fragments of vases in Nippur, stands with *Lugal-si-kisal* on Pl. 37, No. 86, 11 f. e.—1, I am inclined to regard them as father and son. Cf. also No. 89.

[4] Cf. Hilprecht, *Recent Research in Bible Lands,* p. 67.

[5] Cf. already Amiaud in *The Babylonian and Oriental Record* I, pp. 120 ff. On the reading of *Sugir* instead of *Girsu* cf. also Hommel, *Geschichte,* pp. 290, 292, 296, etc., and Jensen, in Schrader's *K. B.* III, part 1, pp. 11 f. (note).

[6] With George Smith, Amiaud, Hommel and others (against Lehmann, *Shamashshumukin,* p. 73). That *Agade* can go over into *Akkad* philologically, I can prove from other examples. But even if this was not the case, the clear statement of George Smith (cf. Delitzsch, *Paradies,* p. 198) should be sufficient. I cannot admit the possibility of a original mistake on the part of George Smith. Master in reading cuneiform tablets as he was, he could not have made a blunder which would scarcely happen to a beginner in Assyriology.

[7] That Akkad became finally identical with "the Babylonian empire in its political totality and unity," was demonstrated by Lehmann, *l. c.,* pp. 71 ff.

TABLE OF CONTENTS

AND DESCRIPTION OF OBJECTS.

Part II, Plates 36–70 and XVI–XXX.

ABBREVIATIONS.

angul., angular; **beginn.**, beginning; **c.**, circa; **ca.**, cast; **C. B. M.**, Catalogue of the Babylonian Museum, University of Pennsylvania (prepared by the editor); **cf.**, confer; **col.**, column(s); **Coll.**, Collection; **d.**, diameter; **Dyn.**, Dynasty; **E.**, East(ern); **f.**, following page; **ff.**, following pages; **f. e.**, from (the) end; **follow.**, following; **fr.** or **fragm.**, fragment(s), fragmentary; **h.**, height; **horizont.**, horizontal; **ibid.**, ibidem; **inscr.**, inscription; **l.** or **li.**, line(s); **m.**, meter; **M. I. O.**, Musée Impérial Ottoman; **N.**, North(ern); **Nippur I, II, III**, etc., refers to the corresponding numbers on Plate XV; **No.**, Number; **Nos.**, Numbers; **Obv.**, Obverse; **omit.**, omitted; **orig.**, original(ly); **p.**, page; **pp.**, pages; **perpend.**, perpendicular; **Pho.**, Photograph; **Pl.**, Plate; **re.** or **resp.**, respectively; **Recueil**, Recueil de travaux relatifs à la philologie et à l'archéologie égyptiennes et assyriennes, edited by G. Maspero; **restor.**, restored; **Rev.**, Reverse; **S.**, South(ern); **sq.**, squeeze; **T.**, Temple of Bêl; **var.**, variants; **vol.**, volume; **W.**, West(ern); **Z.**, Ziqqurratu; **Z. A.**, Zeitschrift für Assyriologie, edited by C. Bezold.

Measurements are given in centimeters, length (height) × width × thickness. Whenever the object varies in size, the largest measurement is given.

The numbers printed on the left, right and lower margins of Plates 36–42 refer to **C. B. M.** and denote the vase fragments used in restoring the cuneiform texts here published. If more than one fragment is quoted, they are arranged according to their relative importance. On fragments placed in parentheses, as a rule less than one or two complete cuneiform characters are preserved. Fragments originally belonging to the same vase are connected by + or + x +, the former indicating that the breaks of fragments thus joined fit closely together, the latter that an unknown piece is wanting between them.

I. AUTOGRAPH REPRODUCTIONS.

PLATE.	TEXT.	DATE.	DESCRIPTION.
36	86	Lugal-kigub-nidudu.	Fragm. of a large vase in serpentine, 20.5 × 9.45 × 2.8, orig. d. c. 25.4. *Nippur* III, beneath the rooms of T. on the S. E. side of Z., a little above Ur-Ninib's pavement in the same stratum as has produced nearly all the fragments of the most ancient stone vases so far excavated in Nuffar (approximately therefore the same place as Pl. 1, No. 1). Inscr. 15 (orig. at least 30) li. C. B. M. 9825. Portions of these 15 li. preserved on the follow. 21 other fragm. of vases in calcite stalagmite (from which the text had been restored before 9825 was found and examined): C. B. M. 9657 + 9607 + 9609 (cf. Pl. XVIII, Nos. 41–43), 9581 + 9643, 9608 + 9679 + 9591 (belonging to the same vase as 9900, cf. Pl. 37 and Pl.

PLATE.	TEXT.	DATE.	DESCRIPTION.
			XVIII, No. 47), 9901, 9902, 9903, 9904 (cf. Pl. 37), 9905, 9632 (belonging to the same vase as 9635 + 9620 + 9627 + 9606, cf. Pl. 37), 9605 (cf. Pl. XVIII, No. 44), 9599, 9633, 9680, 9703, 10001 (cf. Pl. XVIII, No. 48). Cf. also 9634 (cf. Pl. 37 and Pl. XVIII, No. 46).
37	86	Lugal-kigub-nidudu.	The same inscr. continued. On the scale of fr. 9325 restored from 16 fragm. of vases in white calcite stalagmite. *Nippur* III, approximately same place as Pl. 36. C. B. M. 10001 (cf. Pl. 36 and Pl. XVIII, No. 48), 9900 (cf. Pl. XVIII, No. 47, belonging to the same vase as 9608 + 9679 + 9591, cf. Pl. 36), 9904 (cf. Pl. 36), 9620 + 9627 + 9635 + 9606 (belonging to the same vase as 9632, cf. Pl. 36), 9604, 9630, 9631, 9917 (red banded), 9639, 9644. Cf. also 9634 (cf. Pl. 36 and Pl. XVIII, No. 46), 9607 (cf. Pl. 36 and Pl. XVIII, No. 41), 9613 (cf. Pl. XVIII, No. 40).
38	87	Lugalzaggisi.	Five fragm. of a vase in white calcite stalagmite (glued together), 16 × 13 × 1.9. *Nippur* III, approximately same place as Pl. 36, No. 86. Inscr. 3 col., 13 + 17 + 8 = 38 li. C. B. M. 9914 + 9910 + 9915 + 9913 + 9320. Cf. Pl. XIX, No. 49. On the basis of these five fragm. the complete text published on Plates 38–42 has been restored by the aid of the follow. 83 other fragm. belonging to 63 different vases: C. B. M. 8614, 8615, 9300, 9301, 9304, 9306, 9307 + x + 9668, 9308, 9309 + 9924 + 9311 + 9316 + 9314 + 9916, 9312 (cf. Pl. XIX, No. 59), 9317, 9318 + 9645, 9583, 9584 + 9315, 9587, 9595, 9598, 9601 + 9305, 9602, 9611 + x + 9610 (cf. Pl. XIX, Nos. 50, 51), 9619, 9624, 9625, 9628 (cf. Pl. XIX, No. 53), 9638, 9642, 9646 + x + 9310, 9651 + 9911, 9654, 9656 + 9685 (cf. Pl. XIX, No. 58), 9659 + 9660 + 9319, 9662 + 9665, 9663, 9666, 9667, 9670, 9671, 9673, 9674, 9683 (cf. Pl. XIX, No. 60), 9687 (cf. Pl. XIX, No. 61), 9689, 9692 (cf. Pl. XIX, No. 56), 9695 (cf. Pl. XIX, No. 57), 9696 + 9637 (cf. Pl. XIX, No. 52), 9697 + x + 9927, 9698, 9700 (cf. Pl. XIX, No. 55), 9701, 9702, 9903, 9905, 9906, 9907, 9908, 9912 + 9658, 9921 + 9313, 9922, 9923, 9925 (cf. Pl. XIX, No. 54), 9926, 9928, 9929.
39	87	Lugalzaggisi.	The same, continued.
40	87	Lugalzaggisi.	The same, continued.
41	87	Lugalzaggisi.	The same, continued.
42	87	Lugalzaggisi.	The same, continued.
42	88	Lugal-kigub-[nidudu].	Fragm. of a vase in white calcite stalagmite, 2 7 × 10 × 2. *Nippur* III, approximately same place as Pl. 36, No. 86. Inscr. 3 col., 1 + 3 + 2 = 6 li. C. B. M. 9900.
42	89	Lugal-kisalsi.	Two fragm. of a vase in white calcite, probably stalagmite (glued together), 4.85 × 4.9 × 2. *Nippur* III, approximately same place as Pl. 1, No. 1. Inscr. 4 li. C. B. M. 9648 a and b. Cf. Pl. 37, No. 86, li. 7–5 f. e.
42	90	En-shagsag(?)-anna.	Fragm. of a vase in white calcite stalagmite, 5.8 × 7.8 × 1.8. *Nippur* III, approximately same place as Pl. 36, No. 86. Inscr. 5 li. C. B. M. 9930.
43	91	En-shagsag(?) anna.	Two fragm. of a vase in white calcite stalagmite (glued together), 4.8 × 5.5 × 1.2. *Nippur* III, approximately same place as Pl. 36,

PLATE.	TEXT.	DATE.	DESCRIPTION.
			No. 86. Inscr. 3 (orig. 5) li. C. B. M. 9963 + 9998. For the end of the inscr. cf. Pl. 43, No. 92.
43	92	En-shagsag(?)-anna.	Fragm. of a vase in white calcite stalagmite, 4.5 × 9 × 1.6. *Nippur* III, approximately same place as Pl. 1, No. 1. Inscr. 3 (orig. 5) li. C. B. M. 9618. For the beginn. of the inscr. cf. Pl. 43, No. 91.
43	93	Ur-Shulpauddu.	Two fragm. of a vase in white calcite stalagmite (glued together), 12.5 × 6 × 1. *Nippur* III, approximately same place as Pl. 1, No. 1. Inscr. 8 li. C. B. M. 9616 + 9931 (the former excavated 1890, the latter 1893). Parts of li. 2–7 written also on C. B. M. 9622.
43	94	Ur-Enlil.	Votive tablet in impure bluish gray limestone, round hole in the centre, 2 groups of figures and an inscription incised; 20.6 × 19.3 × 2.6, d. of the hole 3.2. *Nippur* X, found out of place in the loose earth along the S. W. side of the Shatt-en-Nil, c. ½ m. below surface. Between the figures of the upper group 4 li. of inscr., beginning on the right, the last 2 li. separated by a line. Sq. Cf. Pl. XVI, No. 37.
43	95	Ur-Mama.	Fragm. of a vase in brownish limestone with veins of white calcite, 5.8 × 6.9 × 1. *Nippur* III, approximately same place as Pl. 1, No. 1. Inscr. 4 (orig. probably 5) li. C. B. M. 9652.
44	96	Aba-Enlil.	Two fragm. of an alabaster bowl (badly decomposed), 12.2 × 7.2 × 1.1. *Nippur* III, approximately same place as Pl. 1, No. 1. Inscr. 10 li. C. B. M. 9621 + 9617.
44	97	[Ur?]-Enlil.	Fragm. of a vase in white calcite stalagmite, 5.1 × 3.3 × 1.4. *Nippur* III, approximately same place as Pl. 36, No. 86. Inscr. 4 li. C. B. M. 9932.
44	98	Same Period.	Two fragm. of a vase in white calcite stalagmite (glued together), 8.4 × 6.9 × 1. *Nippur* III, approximately same place as Pl. 36, No. 86. Inscr. 7 li. C. B. M. 9952 + 9699 (the former excavated 1893, the latter 1890).
44	99	Same Period.	Fragm. of a vase in white calcite stalagmite, 9.7 × 6.3 × 1.6. *Nippur* III, approximately same place as Pl. 36, No. 86. Inscr. 6 li., beginn. of each li. wanting. C. B. M. 9953.
44	100	Same Period.	Fragm. of a vase in white calcite stalagmite, 3.8 × 5.8 × 1.1. *Nippur* III, approximately same place as Pl. 1, No. 1. Inscr. 2 li. C. B. M. 9636.
44	101	Same Period.	Fragm. of a vase in white calcite stalagmite, 4.2 × 4.5 × 0.5. *Nippur* III, approximately same place as Pl. 1, No. 1. Inscr. 3 li. C. B. M. 9686.
45	102	Time of Ur-Shulpauddu.	Fragm. of a vase in white calcite stalagmite, 8.5 × 9.5 × 2.7. *Nippur* III, approximately same place as Pl. 1, No. 1. Inscr. 7 li. C. B. M. 9614. Parts of li. 1–4 written also on C. B. M. 9297 (dark brown sandstone), which apparently belongs to the same vase as Pl. 45, No. 103 and Pl. 46, No. 110.
45	103	Same Period.	Two fragm. of a vase in dark brown sandstone (glued together), 7.6 × 4.3 × 1.3. *Nippur* III, approximately same place as Pl. 36, No. 86. Inscr. 5 li. C. B. M. 9954 + 9924. To the same vase belongs Pl. 46, No. 110. Text supplemented by the follow. two Nos.

PLATE.	TEXT.	DATE	DESCRIPTION.
45	104	Same Period.	Fragm. of a vase in dark brown tufa (decomposed igneous rock), 7.4 × 7.3 × 1. *Nippur* III, approximately same place as Pl. 36, No. 86. Inscr. 7 li. C. B. M. 9951. Text supplemented by Pl. 45, Nos. 103, 105 and Pl. 46, No. 110.
45	105	Same Period.	Fragm. of a vase in dark brown tufa, 5.4 × 4.9 × 0.8. *Nippur* III, approximately same place as Pl. 1, No. 1. Inscr. 5 li. C.B.M. 9623. Text supplemented by Pl. 45, Nos. 103, 104 and Pl. 46, No. 110.
45	106	Same Period.	Two fragm. of a vase in bluish banded calcite stalagmite (glued together), 4.4 × 6.1 × 0.8. *Nippur* III, approximately same place as Pl. 1, No. 1. Inscr. 4 li. C. B. M. 9682 + 9629.
45	107	A patesi (?) of Shirpurla.	Fragm. of a vase in grayish calcite stalagmite, 3.1 × 5.6 × 0.8. *Nippur* III, approximately same place as Pl. 1, No. 1. Inscr. 2 li. C. B. M. 9597.
46	108	A patesi of Kîsh.	Fragm. of a vase in dark brown sandstone, 13.3 × 7.5 × 1.7. *Nippur* III, approximately same place as Pl. 1, No. 1. Inscr. 4 li. C. B. M. 9572. To the same vase belongs the follow. No.
46	109	A patesi of Kîsh.	Two fragm. of the same vase (glued together), 13 × 14.5 × 1.7. *Nippur* III, approximately same place as previous No. Inscr. 4 li. C. B. M. 9571 + 9577.
46	110	Time of Ur-Shulpauddu.	Three fragm. of a vase in dark brown sandstone (glued together), 16.7 × 11 × 1.5. *Nippur* III, approximately same place as Pl. 1, No. 1. Inscr. 9 li. C. B. M. 9574 + 9575 + 9579. To the same vase belongs Pl. 45, No. 103. Text supplemented by Pl. 45, Nos. 104, 105.
47	111	Time of Ur-Enlil.	Two fragm. of a vase in white calcite stalagmite, orig. h. c. 14, d. at the bottom c. 16.5. Fragm. 9302: 9.5 × 8.9 × 1.9. Fragm. 9600: 8.2 × 11.8 × 1.9. *Nippur* III, approximately same place as Pl. 36, No. 86. Inscr. (beginn. and end) 3 + 3 = 6 li. C. B M. 9302, 9600.
47	112	Time of Ur-Shulpauddu.	Fragm. of a vase in bluish banded calcite stalagmite, inside blackened, 13.2 × 15.4 × 2.3, orig. d. 17.4. *Nippur* III, approximately same place as Pl. 36, No. 86. Inscr. 8 × 4.5, 7 li. C. B. M. 9329.
47	113	A little later.	Fragm. of a vase in brownish gray calcite stalagmite, 17.1 × 11 × 1.35, orig. d. at the centre 17.3. *Nippur* III, approximately same place as Pl. 36, No. 86. Inscr. 10 × 3, 13 li. C. B. M. 9330.
47	114	Same Period.	Fragm. of a vase in white calcite stalagmite, 6.8 × 6.5 × 1.1. *Nippur* III, approximately same place as Pl. 1, No. 1. Inscr. 6 li. C. B. M. 9655.
48	115	Entemena.	Two fragm. of a large vase in white calcite stalagmite, outside blackened, 13.4 × 14.8 × 3. *Nippur* III, approximately same place as Pl. 1, No. 1. Inscr. 2 col., 8 + 6 = 14 li. C. B. M. 9163 + 9690 (both excavated 1890). To the same vase belong the follow. two Nos.
48	116	Entemena.	Fragm. of the same vase, 9.4 × 7.2 × 2.7. *Nippur* III, approximately same place as Pl. 36, No. 86. Inscr. 2 col., 4 + 3 = 7 li. C. B. M. 9328 (excavated 1893).
49	117	Entemena.	Two. fragm. of the same vase, 7.1 × 9.9 × 2.6. *Nippur* III, approximately same place as previous No. Inscr. 2 col., 5 + 2 = 7 li. C. B. M. 9919 + 9920 (both excavated 1893).

PLATE.	TEXT.	DATE.	DESCRIPTION.
49	118	Dyn. of Kîsh.	Fragm. of a vase in coarse-grained diorite, 12 × 12.2 × 1.6. *Nippur* III, approximately same place as Pl. 36, No. 86. Inscr. 6 li. C. B. M. 9918.
49	119	Sargon I. (?)	Fragm. of a vase in white calcite stalagmite, 4.8 × 8.4 × 1. *Nippur* III, approximately same place as Pl. 36, No. 86. Inscr. 4 (orig. 6) li. C. B. M. 9331.
50	120	Narâm-Sin.	Fragm. of an inscribed bas-relief in basalt, 52.5 × 39.7 × 8.5. *Diarbekir.* Inscr. 19.1 × 18.4, 4 col., 2 + 6 + 8 + 8 = 24 li. Ca. Orig. M. I. O., Constantinople. Cf. Pl. XXII, No. 64; also Scheil in *Recueil* XV, pp. 62–64, Maspero, *ibid.*, pp. 65f. and *The Dawn of Civilization*, pp. 601f., Hilprecht, *Recent Research in Bible Lands*, pp. 87–89.
51	121	Ur-Gur.	Door socket in a black dense trachytic rock, 41 × 25 × 18. *Nippur* III, 12½ m. below surface, underneath the W. corner of the S. E. buttress of Z. Inscr. 19.7 × 7.5, 10 li. Sq.
52	122	Ur-Gur.	Gray soapstone tablet, Obv. flat, Rev. rounded, 12.2 × 7.7 × 1.7. *Nippur* III, approximately same place as Pl. 36, No. 86. Inscr. 5 li. (identical with that on his bricks). C. B. M. 9932. Cf. I R. 1, No. 9.
52	123	Dungi.	Dark gray soapstone tablet, Obv. flat, Rev. rounded, 8.3 × 5.6 × 1.6. *Nippur* X, found out of place in the rubbish at the foot of a mound, c. 1 m. above the surface of the plain. Inscr. 6 (Obv.) + 2 (Rev.) = 8 li. Sq.
53	124	Dungi.	Fragm. of a baked clay tablet, reddish brown with black spots, Obv. flat, Rev. rounded, 20.1 × 18.5 × 4.3. *Tello.* Obv., 6 col. (23 + 30 + 35 + 22 + 22 + 25 =) 157 li. Orig. in M. I. O., Constantinople (Coll. Rifat Bey, No. 242), copied there 1894. Pl. ⅔ of orig. size.
54	124	Dungi.	The same, Rev., 6 col. (21 + 15 + 10 + 27 + 35 + 18 =) 126 li. Copied in Constantinople 1894. Pl. ⅔ of orig. size.
55	125	Ine-Sin.	Two fragm. of a baked clay tablet, light brown (glued together), Obv. flat, Rev. rounded, 12.8 × 6.1 × 2.8. *Nippur* X. Inscr. 19 (Obv.) + 22 (Rev.) = 41 li. Orig. in M. I. O., Constantinople, copied there 1893. Cf. Hilprecht, *Assyriaca*, pp. 22f., Scheil, in *Recueil* XVII, pp. 37f.
56	126	Bur-Sin II.	Baked clay tablet, reddish brown, Obv. flat, Rev. rounded, 20.5 × 19.9 × 3.8. *Tello.* Obv., 7 col. (parts of col. I–III, VI, VII wanting, 32 + 19 + 32 + 31 + 31 + 30 + 21 =) 196 li. Orig. in M. I. O., Constantinople (Coll. Rifat Bey, No. 256), copied there 1894. Pl. ⅔ of orig. size.
57	126	Bur-Sin II.	The same, Rev., 7 col. (part of col. I wanting, 30 + 23 + 21 + 20 + 23 + 15 + 10 =) 142 li. Copied in Constantinople 1894. Pl. ⅔ of orig. size.
58	127	Gimil (Kât)-Sin.	Fragm. of a clay tablet, slightly baked, dark brown, Obv. flat, Rev. rounded, 7 × 5 × 2. *Nippur* X. Inscr. 9 (Obv.) + 4 (Rev.) = 13 li. C. B. M.
58	128	Rim-Aku.	Fragm. of a baked clay phallus, light brown, h. 14.3, largest circumference 14.7. *Nippur* X. Inscr. 17 li. Orig. in M. I. O., Constantinople, copied there 1893.

PLATE.	TEXT.	DATE.	DESCRIPTION.
59	129	Ammizaduga.	Two fragm. of a clay tablet, slightly baked, brown, 11.6 × 10.8 × 3.2. *Nippur* X. Obv., 3 col. of inscr., middle col. Sumerian in Old Babylonian characters, first and third col. Semitic Babylonian in Neo-Babylonian script, Rev. badly damaged, traces of second and third col. The tablet was written c. 600 B.C. Orig. in M. I. O., Constantinople.
60	130	Cassite Dyn.	Fragm. of a slab in white marble with reddish veins, 24.5 × 21 × 6.7. *Nippur* III, approximately same place as Pl. 36, No. 86. Inscr. 2 col., 6 + 5 = 11 li. Ca. (C. B. M. 9794). Orig. in M. I. O., Constantinople.
60	131	c. 2500 B.C.	Brown hematite weight, ellipsoidal and symmetrical, complete, weight 85.5 grams, length 7.3, d. 2.1. *Nippur* X (June, 1895). Inscr. 1.9 × 1.8, 3 li. (1. *X shiḳlu* 2. *din ḫurâṣi* 3. *dam-kar* = "10 shekels, gold standard of merchants;" according to this standard 1 mana = 513 gr.). Sq., sent from the ruins.
60	132	Burnaburiash.	Seal cylinder in white chalcedony, length 3.4, d. 1.5. *Babylonia*, place unknown. A bearded standing figure in a long robe, one hand across the breast, the other lifted. A border line at the top. Inscr. 9 li. Impression on gutta percha (in possession of the editor). Orig. in the Metropolitan Museum of Art, New York. Cf. Hilprecht, *Assyriaca*, p. 93, note, Ward, *Seal Cylinders and other Oriental Seals* (Handbook No. 12 of the Metropol. Mus.), No. 391.
60	133	Kurigalzu.	Fragm. of a lapis lazuli disc, 3 2 × 3. *Nippur* X, found in the loose débris on the slope of a mound, and near to its summit (1895). Inscr. 6 (Obv.) + 6 (Rev.) = 12 li. Pencil rubbing, sent from the ruins.
61	134	[Ku]rigalzu.	Fragm. of an agate cameo, 3.95 × 1. *Nippur* III, same place as Pl. 8, No. 15. Inscr. 3 li. Orig. in M. I. O., Constantinople, copied there 1893.
61	135	Kurigalzu.	Fragm. of an agate cameo, 2.8 × 1. *Nippur* III, same place as Pl. 8, No. 15. Inscr. 3 li. Orig. in M. I. O., Constantinople, copied there 1893.
61	136	[Nazi]-Maruttash.	Fragm. of an axe in imitation of lapis lazuli, 6.75 × 4.25 × 1.5. *Nippur* III, same place as Pl. 8, No. 15. Inscr. 7 li. Orig. in M. I. O., Constantinople, copied there 1893. To the same axe belongs the follow. No.
61	137	Nazi-Maruttash.	Fragm. of the same axe, 4.2 × 3.6 × 1.1. *Nippur* III, same place as Pl. 8, No. 15. Inscr. 4 li. Orig. in M. I. O., Constantinople, copied there 1893.
61	138	[Kadashman]-Turgu.	Lapis lazuli disc, 2.75 × 0.3. *Nippur* III, same place as Pl. 8, No. 15, Inscr. of 5 li. (1. [A-na]^{ilu}Nusku 2. be-lī-shú 3. [Ka-dash-man]-Tur-gu 4. a-[na ba]-l [a-ṭi-sh]ú 5. i-[ḳi]-ish) erased in order to use the material. Orig. in M. I. O., Constantinople, copied there 1893.
61	139	Cassite Dyn.	Agate cameo, hole bored parallel with the li., 2.4 × 1.65 × 0.8. *Nippur* III, same place as Pl. 8, No. 15. Inscr. ^{Dingir}En-lil. Orig. in M. I. O., Constantinople, copied there 1893.

PLATE.	TEXT.	DATE.	DESCRIPTION.
61	140	Cassite Dyn.	Remnant of a lapis lazuli tablet the material of which had been used, 2.1 × 2.2. *Nippur* III, same place as Pl. 8, No. 15. Inscr. 3 li. Orig. in M. I. O., Constantinople, copied there 1893.
61	141	Cassite Dyn.	Lapis lazuli disc, 1.2 × 0.15. *Nippur* III, same place as Pl. 8, No. 15. Inscr. *Dingir Nin-lil.* Orig. in M. I. O., Constantinople, copied there 1893.
61	142	Cassite Dyn.	Lapis lazuli disc, 1.2 × 0.15. *Nippur* III, same place as Pl. 8, No. 15. Inscr. *Dingir En-lil.* Orig. in M. I. O., Constantinople, copied there 1893.
61	143	Cassite Dyn. (?)	Fragm. of a light black stone tablet, 2.15 × 2.4 × 0.5. *Nippur* III, same place as Pl. 8, No. 15. Obv., meaning of characters unknown, Rev., animal rampant. Probably used as a charm. Orig. in M. I. O., Constantinople, copied there 1893. Cf. Loftus, *Travels and Researches,* p. 236f.
62	144	Cassite Dyn.	Unbaked clay tablet, dark brown, Obv., nearly flat, Rev., rounded, 6.15 × 4.75 × 1.8. *Nippur* X. Plan of an estate. Orig. in M. I. O., Constantinople, copied there 1893. Cf. Scheil in *Recueil* XVI, pp. 36f.
62	145	Cassite Dyn.	Fragm. of an unbaked clay tablet, dark brown, Obv. nearly flat, Rev. rounded, 3.8 × 6 × 2.35. *Nippur* X. Plan of an estate. C. B. M. 5135.
63	146	Cassite Dyn.	Six fragm. of a slightly baked clay tablet, brown (glued together), Obv. flat, Rev. rounded, 16.5 × 10.5 × 3. *Nippur* X. Inscr., Obv., 4 col., 39 + 40 + 43 + 15 = 137 li., Rev. uninscribed. Orig. in M. I. O., Constantinople, copied there 1894.
64	147	c. 1400 B.C.	Baked clay tablet, dark brown, nearly flat on both sides, upper left corner wanting, 5.9 × 5.2 × 1.6. *Tell el-Hesy* (Palestine), found by F. J. Bliss, at the N. E. quarter of City III, on May 14, 1892. Inscr. 11 (Obv.) + 2 (lower edge) + 11 (Rev.) + 1 (upper edge) + 1 (left edge) = 23 li., irregularly written. Orig. in M. I. O., Constantinople, copied there 1893. Cf. Pl. XXIV, Nos. 66, 67; also Bliss, *A Mound of Many Cities,* pp. 52–60; Sayce, in Bliss's book, pp. 184–187, Scheil in *Recueil* XV, pp. 137f., Conder, *The Tell Amarna Tablets,* pp. 130–134 (worthless!).
64	148	Marduk-shâbik-zêrim.	Fragm. of a baked clay cylinder, barrel shaped, solid, light brown; h. of fragm. 7.98, orig. d. at the top c. 5.3, at the centre c. 7.8. *Place unknown.* Inscr. 2 (orig. 4) col., 16 + 22 + 1 (margin) = 39 li. Orig. in possession of Dr. Talcott Williams, Philadelphia, Pa. Cf. Pl. XXIV, No. 68; also Jastrow, Jr., in *Z. A.* IV, pp. 301–325, VIII, pp. 214–219, Knudtzon, *ibid.,* VI, pp. 163–165, Hilprecht, *ibid.,* VIII, pp. 116–120, and Part I of the present work, p. 44, note 4.
65	149	Marduk-aḫê-irba.	Boundary stone in grayish limestone, irregular, 48.5 × 24.5 × 18. *Babylonia,* place unknown. Figures facing the right. Upper section: Turtle (on the top of the stone); scorpion, crescent, disc of the sun, Venus (all in the first row below); 2 animal heads with long necks (cf. V R. 57, sect. 4, fig. 1), bird on a post, object similar to V R. 57, sect. 2, with an animal resting alongside (sim-

PLATE.	TEXT.	DATE.	DESCRIPTION.
			ilar to V R. 57, sect. 3, fig. 1), same object without animal (all in the second row below) ; object similar to V R. 57, sect. 6, but without animal (below the 2 animal heads). Lower section : A seated figure, both hands lifted (cf. V R. 57, sect. 5, fig. 1), object similar to V R. 57, sect. 6, last object, but reversed, large snake. Inscr. 3 col., 22 + 23 + 11 = 56 li. Sq. Orig. in private possession, Constantinople. Cf. Hilprecht, *Assyriaca*, p. 33, Scheil in *Recueil* XVI, pp. 32f. Pl. $\frac{2}{8}$ of orig. size.
66	149	Marduk-aḫê-irba.	The same, continued. Pl. $\frac{2}{8}$ of orig. size.
67	149	Marduk-aḫê-irba.	The same, continued. Pl. $\frac{2}{8}$ of orig. size.
68	150	c. 1100 B.C.	Upper part of a black boundary stone, 33 × 38 × 20. *Nippur*. Inscr. 2 col., 6 + 6 = 12 li. Ca. Orig. in the Royal Museums, Berlin. Cf. Pl. XXV, No. 69; also *Verzeichniss der (in den Königlichen Museen zu Berlin befindlichen) Vorderasiatischen Altertümer und Gipsabgüsse*, p. 66, No. 213.
69	151	Esarhaddon.	Fragm. of a baked brick, yellowish, partly covered with bitumen, 18.5 (fragm.) × 7.3 (fragm.) × 8 (orig.). *Babylon*. Inscr. (written on the edge) 15 × 6, 11 li. C. B. M. 14.
70	152	Nebuchadrezzar II.	Fragm. of a baked brick from the outer course of a column, 22.2 (fragm.) × 35 (orig.) × 9.2 (orig.). *Abu Habba*. Inscr. (written on the outer surface) 33.6 × 8, 3 col., 8 + 8 + 8 = 24 li. Sq. Orig. in M. I. O., Constantinople.

II. Photograph (half-tone) Reproductions.

XVI	37	Ur-Enlil.	Votive tablet in impure bluish gray limestone, figures and inscription incised. *Nippur*. Upper section : A naked (uncircumcised) worshiper (Ur-Enlil) standing before a seated god and offering a libation. Same group reversed on the left. Between the figures 4 li. of inscr. Lower section : A goat and a sheep followed by two men, one carrying a vessel on his head, the other holding a stick in his right hand. Pho. taken from a sq. Cf. Pl. 43, No. 94.
XVI	38	Same Period.	Two fragm. of a votive tablet in impure bluish gray limestone, round hole in the centre, figures incised, 17.2 × 18.6 × 3, d. of the hole 1.7. *Nippur* III, found out of place, in the débris filling one of the rooms of T. to the S. W. of Z., not far below surface. Upper section : A naked worshiper standing before a seated god and offering a libation. The god reversed on the left. Lower section : A gazel walking by a bush (or nibbling at it ?), a hunter about to draw his bow at her. Orig. in M. I, O., Constantinople. Pho. taken from a ca. (C. B. M. 4984).
XVII	39	Lugal-kigub-nidudu.	Unhewn block of white calcite stalagmite, 29 × 21 × 19.5. *Nippur* III, c. 10 m. below surface under the rooms of T. on the S. E. side of Z. Inscr. 10.3 × 6, 4 (orig. 8 ?) li. C. B. M. 10050.

PLATE.	TEXT.	DATE.	DESCRIPTION.
XVIII	40–48	Lugal-kigub-nidudu.	Fragm. of vases in white calcite stalagmite, from which (together with others) the text on Plates 36, 37 has been restored. *Nippur.* C. B. M. 9613, 9607 + 9657 + 9609, 9605, 9634, 9900, 9606, 10001. Cf. Plates 36, 37, No. 86.
XIX	49–61	Lugalzaggisi.	Fragm. of vases in white calcite stalagmite, from which (together with others) the text on Plates 38–42 has been restored. *Nippur.* C. B. M. 9914 + 9910 + 9915 + 9913 + 9320, 9611 + x + 9610, 9696 + 9637, 9628, 9925, 9700, 9692, 9695, 9685, 9312, 9683, 9687. Cf. Plates 38–42, No. 87.
XX	62	Al-usharshid.	White marble vase, an inscribed portion (containing parts of li. 8, 9, 11–13 and the whole of li. 10) broken from its side. *Nippur* III, approximately same place as Pl. 36, 37, No. 86. Inscr. 20.6 × 5.6, 13 li. Orig. in M. I. O., Constantinople. Pho. taken from a ca. (C. B. M. 9793). Cf. Pl. 4, No. 5 and Pl. III, Nos. 4–12.
XXI	63	Sargon I.	Fragm. of a brick of baked clay, yellowish, 23.5 (fragm.) × 18 (fragm.) × 8 (orig.). *Nippur* III, found out of place on the S. E. side of Z., approximately at the same depth as Pl. 36, No. 86. Inscr. (written) 3 li. (orig. 2 col., 6 li.). The character *Shar* repeated on the upper left corner of inscribed surface. Orig. in M. I. O., Constantinople. Cf. Pl. 3, No. 3.
XXII	64	Narâm-Sin.	Fragm. of an inscribed bas-relief in basalt. *Diarbekir.* A god standing on the right, clad in a hairy garment, wearing a conical head-dress. Hair arranged in a net, long pointed beard, bracelets on both wrists, short staff (?) in each hand. Part of hair, left upper arm and both legs wanting. Pho. taken from a ca. (C. B. M. 9479). Cf. Pl. 50, No. 120.
XXIII	65	Ur-Ninib.	Brick of baked clay, light brown, broken, 31 × 15 × 7. *Nippur* III, c. 10 m. below surface underneath the S. E. buttress of Z. from a pavement constructed by Ur-Ninib. Inscr. (written) 22.4 × 10, 13 li., beginning at the bottom. Orig. in M. I. O., Constantinople. Cf. Pl. 10, No. 18.
XXIV	66, 67	c. 1400 B.C.	Tablet of baked clay, Obv. and Rev. *Tell el-Hesy* (Palestine). Pho. taken from a ca. (in possession of the editor). Cf. Pl. 64, No. 147.
XXIV	68	Marduk-shâbik-zêrim.	Fragm. of a baked clay cylinder, barrel shaped, solid, light brown. *Place unknown.* Pho. taken from a ca. (C. B. M. 9553). Cf. Pl. 64, No. 148.
XXV	69	c. 1100 B.C.	Upper part of a black boundary stone. *Nippur.* Upper section: Disc of the sun, crescent, Venus. Lower section: 2 col. of inscr. Pho. taken from a ca. (in possession of the editor). Cf. Pl. 68, No. 150.
XXV	70	Unknown.	Brown sandstone pebble (weight?), oblong, flat on both ends, weight 1067 grams, 8.2 × 14.7 × 6. *Nippur*, on S. E. side of Z., 2¼ m. below surface. Meaning of characters inscribed on convex surface not certain, possibly "⅔ of a mine + 15" = 55 shekels (equal to c. 1054 grams, if referring to the Babylonian heavy silver mine [royal norm = 1146.1–1150.1 gr., according to

PLATE.	TEXT.	DATE.	DESCRIPTION.
			Lehmann in *Actes du huitième congrès international des orientalists*, 1889, Semitic section B, p. 206]). C. B. M. 10049.
XXVI	71	c. 350 B.C.	Bas-relief in baked clay, brown, upper corner and part of lower left corner wanting, 14.3 × 17 × 3.7. *Nippur* III, approximately same place as Pl. XVI, No. 38. Man fighting a lion. Bearded man with a conical head-dress and mass of locks falling over his neck, clad in a short, tight, sleeveless, fringed coat, his left knee resting on the ground. He is thrusting his sword into the flank of a lion, at the same time in defense raising his left arm against the lion's head. The lion, having received a wound over his right foreleg, stands on his hind legs, clutching the sides of his enemy with his fore paws and burying his teeth in the man's left shoulder. Part of man's left foot and of lion's tail and left hind leg wanting. On right side of plinth (0.6 deep) traces of five Aramaic letters, left side broken off. Orig. in M. I. O., Constantinople. Pho. taken from a ca. (C. B. M. 9477).
XXVII	72	At least 4000 B.C.	Terra-cotta vase with rope pattern, in upright position as found in trench, an Arab on each side; h. 63.5, d. at the top 53. *Nippur* III, 5.49 m. below the E. foundation of Ur-Gur's Z.
XXVIII	73	At least 4000 B.C.	Arch of baked brick, laid in clay mortar, h. 71, span 51, rise 33. Bricks convex on one side, flat on the other. Front of arch opened to let light pass through. *Nippur* III, at the orifice of an open drain c. 7 m. below the E. corner of Ur-Gur's Z. View taken from inside the drain.
XXIX	74	Ur-Gur.	N. W. façade of the first stage of Ur-Gur's Z. A section of the drain which surrounded Z. is seen at the bottom of the trench. *Nippur* III.
XXX	75	1894 A.D.	General and distant view of the excavations at T., taken from an immense heap of excavated earth to the E. of Z. *Nippur* III.

CUNEIFORM

TEXTS.

Pl. 37

86
Continued

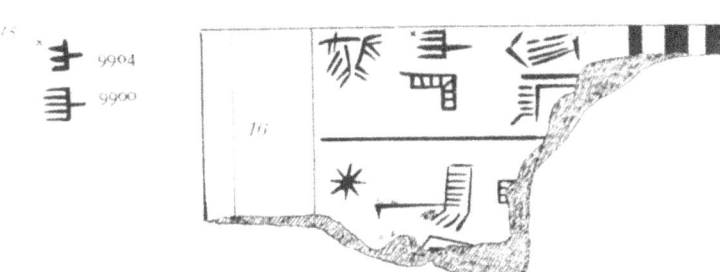

Several lines wanting.

9 f. e. 9635

8 f. e. 9635
.... 9627

7 f. e. 9627

6 f. e. 9627
.... 9630
.... 9630

5 f. e x
.... 9630

l. 16-17 : 10001 for
l. 16 cf. also 9900,
9904.

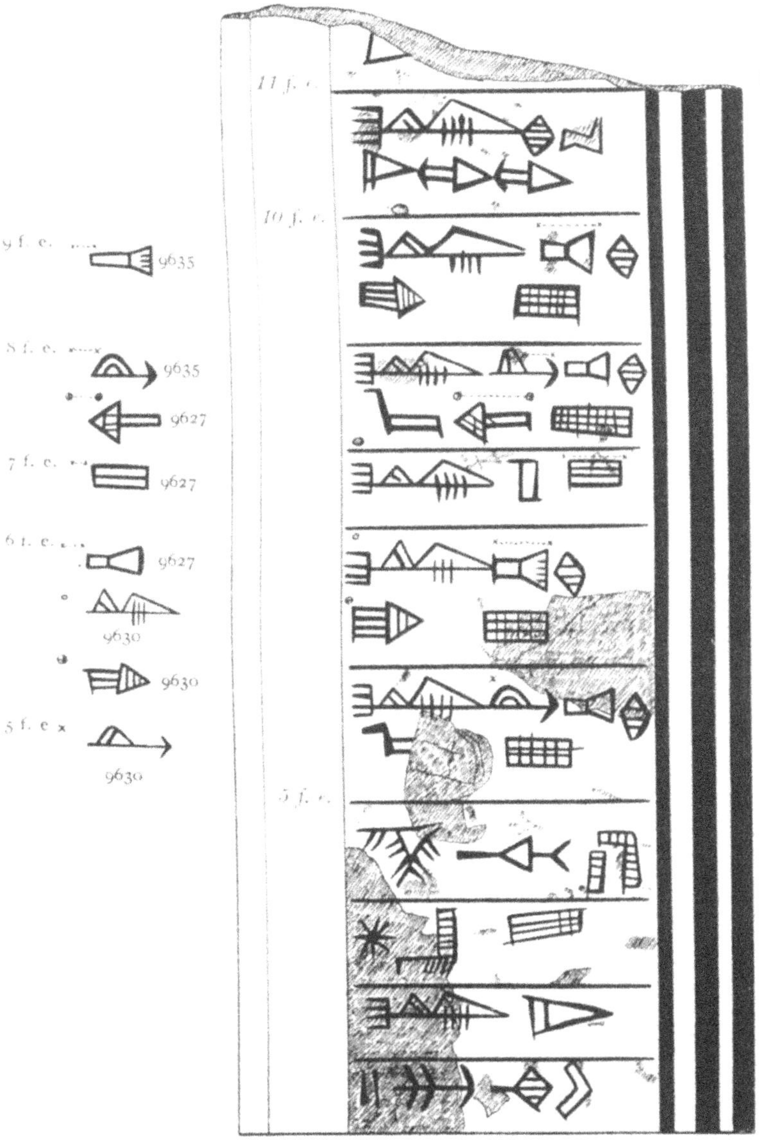

l. 11 f. e. : 9635.

10 f. e. : 9635 - 9620.

9 f. e. : 9620, (9635).

8 f. e. : 9620 - 9627
- 9635 - 9606.

7 f. e. : 9606, 9627,
(9604).

6 f. e. : 9606, 9630,
9627, (9604).

5 f. e. : 9604, (9630,
9631, 9606, 9917
9639).

4-1 f. e. : 9604, beginn.
of l. 3-1 restor. from
9644, for l. 4 cf.
(9631, 9639, 9634,
9917).

2 f. e. : (9917, 9639).

1 f. e. : (9607).

Trans. Am. Phil. Soc., N. S. XVIII, 8.

Pl. 38

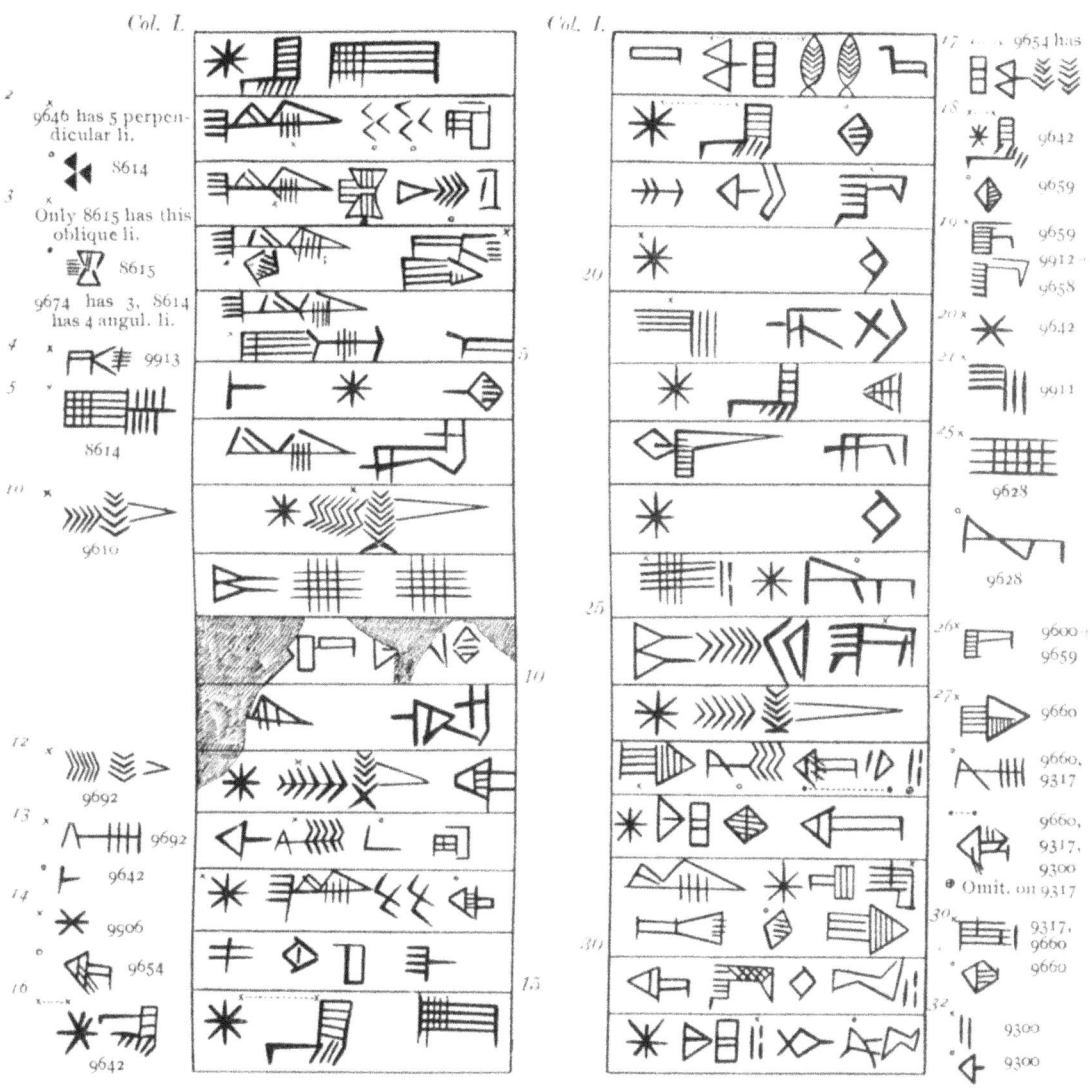

NOTE.—The above text has been restored from the following fragments, COL. 1, L. 1: frr. 8614, 9646, (9313, 9915, 9611, 9923). L. 2: 8614, 8615, 9646, 9921+9313, 9115+9913, 9611, (9674, 9923). L. 3: 8614, 8615, 9913, 9674, 9662, (9313). L. 4: 8614, 8615, 9674, 9913, 9662, (9587). L. 5: 8614, 8615, 9674, 9913, (9662, 9587). L. 6: 8615, 9610, (9913, 9674, 9587). L. 7: 8615, 9610, (9587). L. 8-9: Ibidem. L. 10: (9692, 9642). L. 11: 9696, (9692, 9642, 9689). L. 12: 9696+9637, 9642, 9692, (9689). L. 13: 9642, 9637, 9689, 9583, (9692, 9654, 9906). L. 14: 9642, 9654, (9689, 9583, 9906, 9637). L. 15: 9642, 9654, 9318, 9583, 9906, (9689, 9656). L. 16: 9642, 9318, 9654, 9906, (9583, 9689, 9656, 9659+9319). L. 17: 9318, 9642, 9654, 9906, (9912+9658, 9583, 9659+9319). L. 18: 9318, 9642, [written on L. 17], 9906, (9912+9658, 9654, 9659). L. 19: 9318, 9642, (9317, 9651, 9912+9658, 9702, 9659, 9906). L. 20: 9317, 9318, 9651, (9642, 9702, 9906). L. 21: 9317, 9911+9651, 9645, (9659). L. 22: 9317, 9911, 9645, (9659, 9700). L. 23: 9317, 9645, 9659, (9628, 9700). L. 24: 9317, 9645, 9628, 9659. L. 25: 9317, 9645, 9628, 9659+9660. L. 26: 9317, 9660+9659, (9584, 9645, 9300, 9301). L. 27: 9317, 9660, 9584+9315, 9301, (9300). L. 28: 9584+9315, 9660, 9317, 9301, (9300). L. 29: 9584+9315, 9317, 9301, 9660, (9300, 9307). L. 30: 9584+9315, 9301, 9317, 9660, 9307, 9300. L. 31: 9301, 9584+9315, 9660, 9307, 9300. L. 32: 9301, 9300, (9307, 9315, 9907).

Trans. Am. Phil. Soc., N. S. XVIII, 8.

Pl. 39

87
Continued

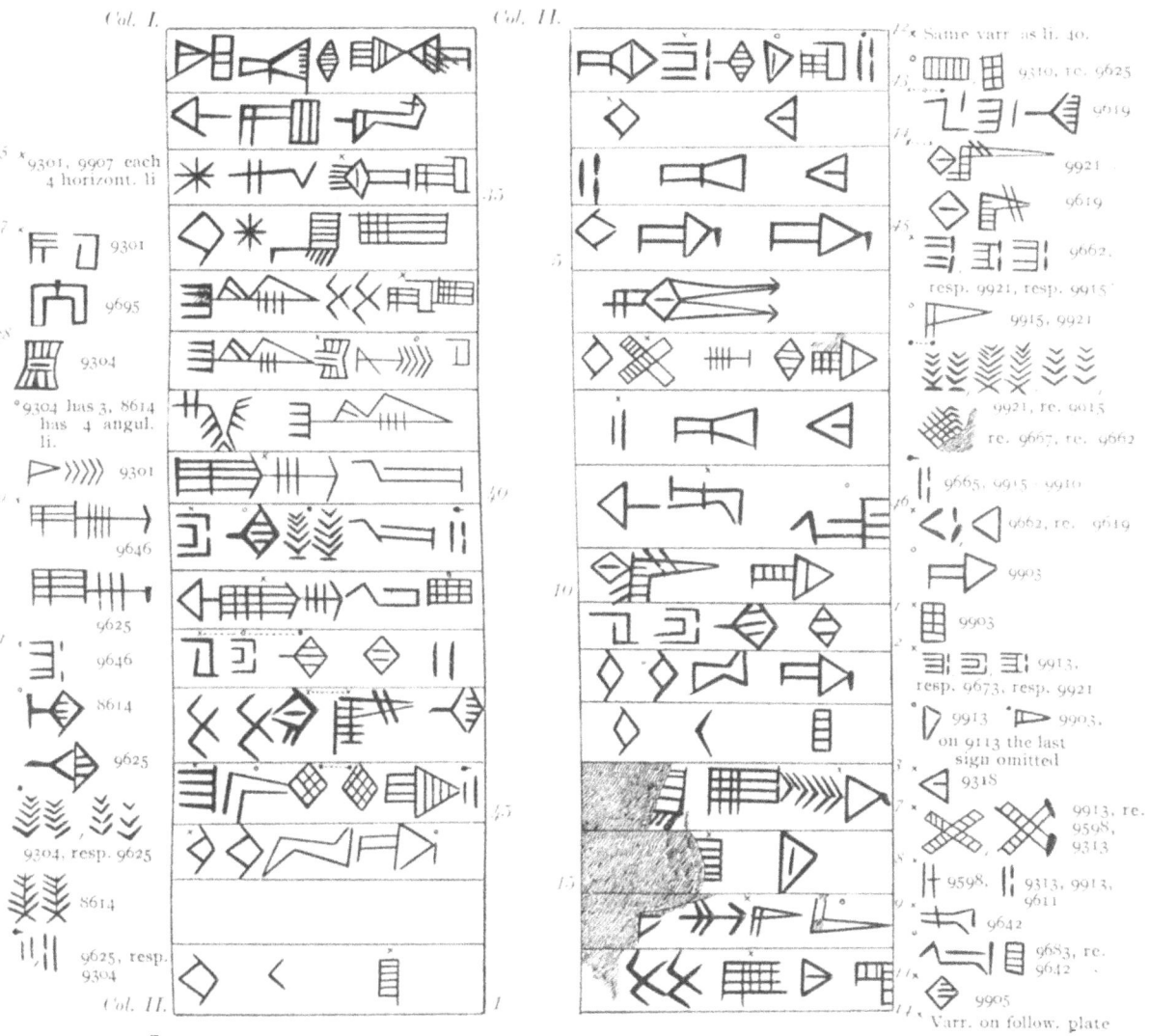

L. 33: 9907, 9301, 8614, 9300, (9306). L. 34: 9301, 8614, 9907, (9306). L. 35: 9301, 8614, 9907, 9306. L. 36: 9301, 8614 [col. II begins], 9306, (9907, 9695). L. 37: 8614, 9301, 9306, (9695, 9304). L. 38: 8614, 9301, 9304, 9306, (9695, 9646). L. 39: 8614, 9304, 9646, 9625, 9306, (9595, 9695, 9638). L. 40: 8614, 9304, 9646, 9625, 9638, 9306, (9695, 9914). L. 41: 8614, 9304, 9646 [col. I ends], 9625, 9306, (9914, 9638, 9695). L. 42: 9304, 8614, 9619, 9625, 9306 [col. I ends], 9310 [col. II begins], (9914, 9921). L. 43: 9619, 9304, 9662, 9701, (9921, 9914+9910, 9310). L. 44: 9619, 9662+9665, 9915+9910, 9921, 9701, (9922). L. 45: 9619, 9915+9910, 9662+9665, 9921, (9667, 9922). L. 46: 9921, 9619, 9915, 9667, (9908, 9665, 9922, 9318, 9662). COL. II, L. 1: 9913, 9921, 9667, 9903, (9318, 9662). L. 2: 9921+9313, 9667, 9913, 9903, 9673, (9318). L. 3: 9921, 9667, 9913, 9903, 9673, 9658, (9318). L. 4: 9913, 9313 [col. II begins], 9658, 9903, 9673, (9667). L. 5: 9913, 9313, 9658, 9903, (9673, 9667). L. 6: 9913, 9313, 9658, 9642, (9903, 9645). L. 7: 9313, 9642, (9611, 9913, 9598). L. 8: 9313, 9611, 9642, (9598, 9913, 9683). L. 9: 9611 [col. II begins], 9642, 9905, (9683, 9598, 9313). L. 10: 9611, 9642, (9683, 9905, 9598, 8615, 9674). L. 11: 9611, 9642, 9683, (9905, 9674, 8615). L. 12: 9611, 9642, (9905, 9683, 9674, 8615). L. 13: 9611, 9687, (9642, 9674, 9683, 9905). L. 14: 9905, 9687, (9611, 9671). L. 15: 9305 [col. II begins], (9905, 9671, 9687, 9624). L. 16: 9305, 9624, (9671, 9905). L. 17: 9624, 9610, 9305, (9300).

87
Continued

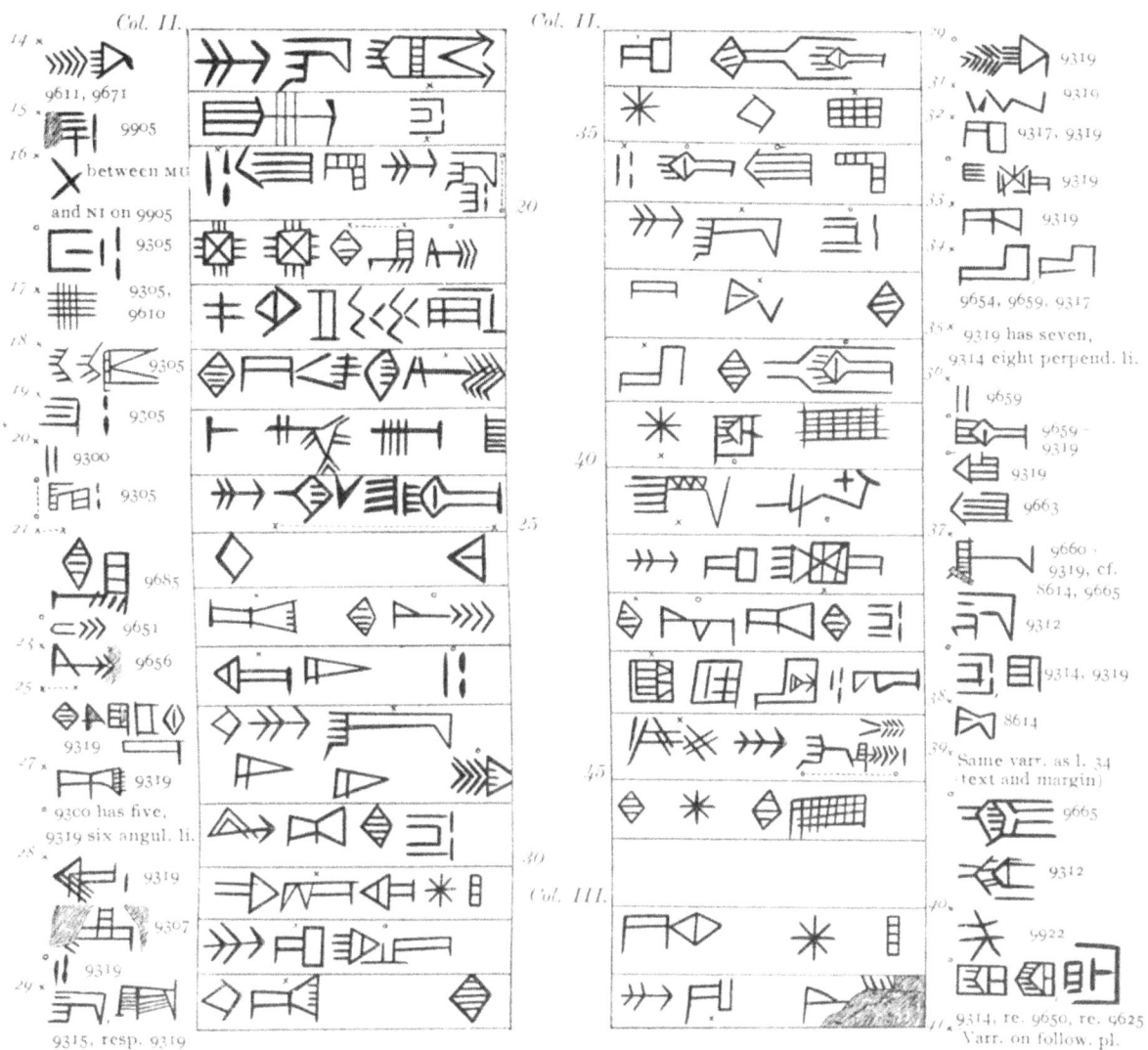

L. **18**: 9610, 9624, 9300, 9305, (9668).   L. **19**: 9610, 9300 [includes the first three characters of L. 20], 9305, (9624).
L. **20**: 9610, 9300, 9305, (9651, 9308, 9685, 9668).   L. **21**: 9610, 9651, 9300, 9685, (9305, 9668, 9308).   L. **22**: 9300, 9651,
9610, 9656, (9319, 9305, 9308).   L. **23**: 9300, 9319, 9656, (9651, 9610).   L. **24**: 9300, 9319, 9656, 9925).   L. **25**: 9300,
9319, (9309, 9315, 9925).   L. **26**: 9300, 9319, 9315, (9309, 9925).   L. **27**: 9319, 9300, 9315, (9309, 9925).   L. **28**: 9319,
9315, (9307, 9309, 9300, 9317).   L. **29**: 9319, 9307, 9315, (9317, 9309).   L. **30**: 9319, 9307, (9315, 9317, 9309).   L. **31**:
9659+9319, 9307, (9317, 9315, 9309, 9654).   L. **32**: 9307, 9659+9319, 9317, 9654.   L. **33**: 9307, 9659+9319, 9654, 9317,
(9907, 9314).   L. **34**: 9307, 9659+9319, 9654, 9907, (9317, 9314).   L. **35**: 9307, 9659+9319, 9654, 9907, 9314, (9317, 9663).
L. **36**: 9659+9319, 9307, 8614, 9654, 9907, 9314, (9663, 9317).   L. **37**: 9307, 9660+9659+9319, 8614, 9665, 9314, 9312,
(9654, 9663).   L. **38**: 9307, 8614, 9660+9319, 9665, 9314, 9312, (9914, 9663, 9667).   L. **39**: 8614, 9665, 9307, 9660+9319,
9914, 9314, 9312, (9922, 9667, 9625).   L. **40**: 8614 [col. III begins], 9665, 9914, 9307, 9625, 9660, 9314, (9922, 9667).   L. **41**:
9914, 8614, 9660, 9665, 9314, (9625, 9922, 9307).   L. **42**: 9914+9320, 8614, 9314+9316, (9660, 9665, 9922).   L. **43**: 9914+
9320, 8614, 9314+9316, (9646+x+9310, 9922, 9673).   L. **44**: 9910+9914+9320, 8614, 9314+9316, (9310 [col. III begins],
9673, 9922).   L. **45**: 9915+9910+9320, 8614, 9316, (9310).   L. **46**: 9915+9910+9320, 8614, 9316, (9310, 9928).   **Col. III,**
L. **1**: 9913+9320, 9928, 9316, (9903, 8614).   L. **2**: 9913+9320, 9903, 9916+9316, (9928).

Pl. 41

87
Continued

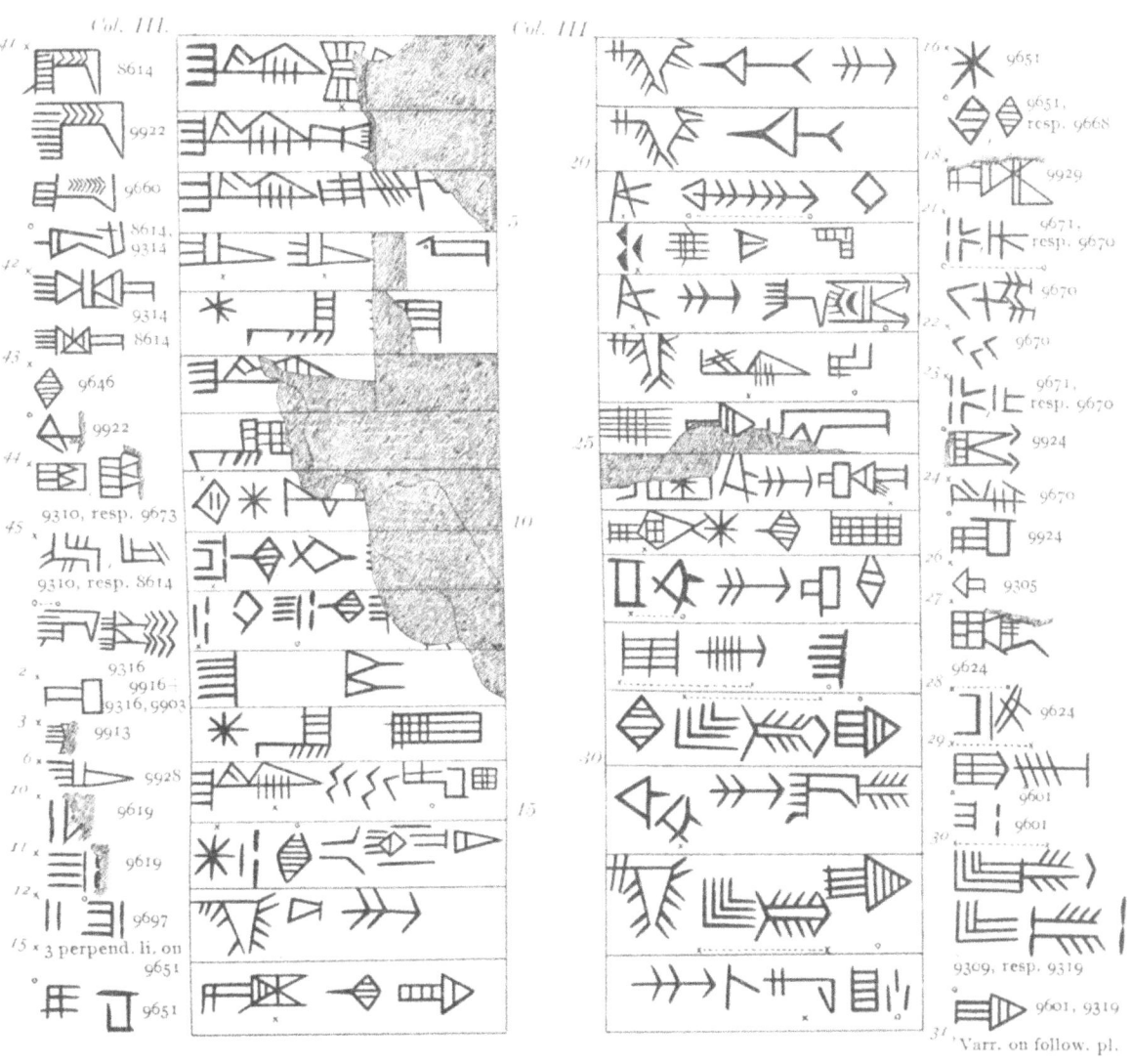

L. 3: 9916+9316, 9903, (9913, 9928). **L. 4:** 9903, 9913, (9928, 9926, 9916). **L. 5:** 5903, 9926, (9928, 9913, 9304). **L. 6:** 9903, 9928, (9926, 9913, 9304). **L. 7:** 9903, (9928, 9304, 9926). **L. 8:** (9304, 9903, 9928). **L. 9:** (9304, 9619). **L. 10:** 9304, (9308, 9619, 9313). **L. 11:** 9308, (9697, 9619, 9313). **L. 12:** 9308, 9697, (9313, 9619). **L. 13:** 9308. **L. 14:** 9308. **L. 15:** 9308, 9651, (9668). **L. 16:** 9308, 9651, (9698). **L. 17:** 9308, (9668, 9924). **L. 18:** 9308, (9929, 9927, 9668, 9924). **L. 19:** 9308, 9929, (9666, 9927, 9924). **L. 20:** 9666, 9929, 9308, (9927, 9924). **L. 21:** 9666, 9670, (9924, 9927, 9671, 9929). **L. 22:** 9666, 9670, (9671, 9924). **L. 23:** 9666, 9670, (9671, 9924). **L. 24:** 9666, 9670, (9671, 9924). **L. 25:** (9666, 9671, 9670, 9305, 9924). **L. 26:** 9305, (9309+9924, 9624). **L. 27:** 9309+9924, 9305 [col. II ends], (9624, 9610). **L. 28:** 9601, 9309+x+9924, 9624, (9663, 9319, 9638, 9610). **L. 29:** 9319, 9309+x+9924, 9601, 9663, (9665, 9624). **L. 30:** 9601, 9663, 9319, 9309, (5665). **L. 31:** 9601, 9663, 9319, 9309, (9665, 9312, 9307). **L. 32:** 9601+9305, 9663, 9319, (9309+9311, 9665, 9312, 9307). **L. 33:** 9305, 9319, 9309+9311, (9665, 9907, 9663).

Trans. Am. Phil. Soc., N. S. XVIII, 8.

Pl. 42

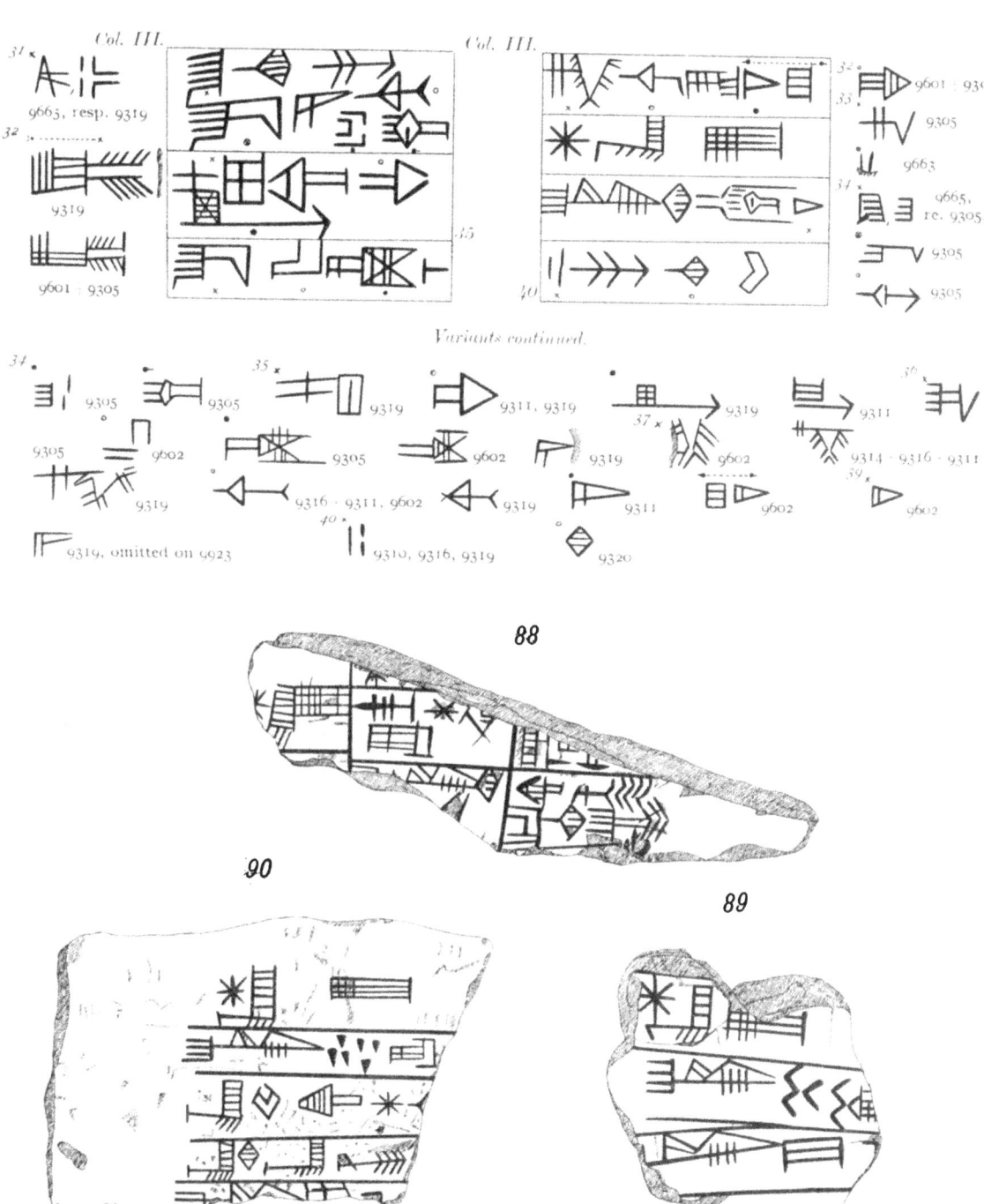

L. 34: 9305, 9319, 9311, (9665, 9307, 8614). L. 35: 9305, 9319, 9316+9311, 8614 [col. III ends], (9602, 9307.) L. 36: 9305, 9314+9316+9311, 9319, 9602, (9307). L. 37: 9305, 9602, 9314+9316+9311, 9319, (9310, 9307). L. 38: 9305, 9602, 9319, 9310, 9314+9316+9311+9923. L. 39: 9305, 9602, 9316+9923, 9319, 9310, (9320). L. 40: 9305, 9316+9923, 9602, 9310, 9320, 9319.

Trans. Am. Phil. Soc., N. S. XVIII, 8.

Pl. 43

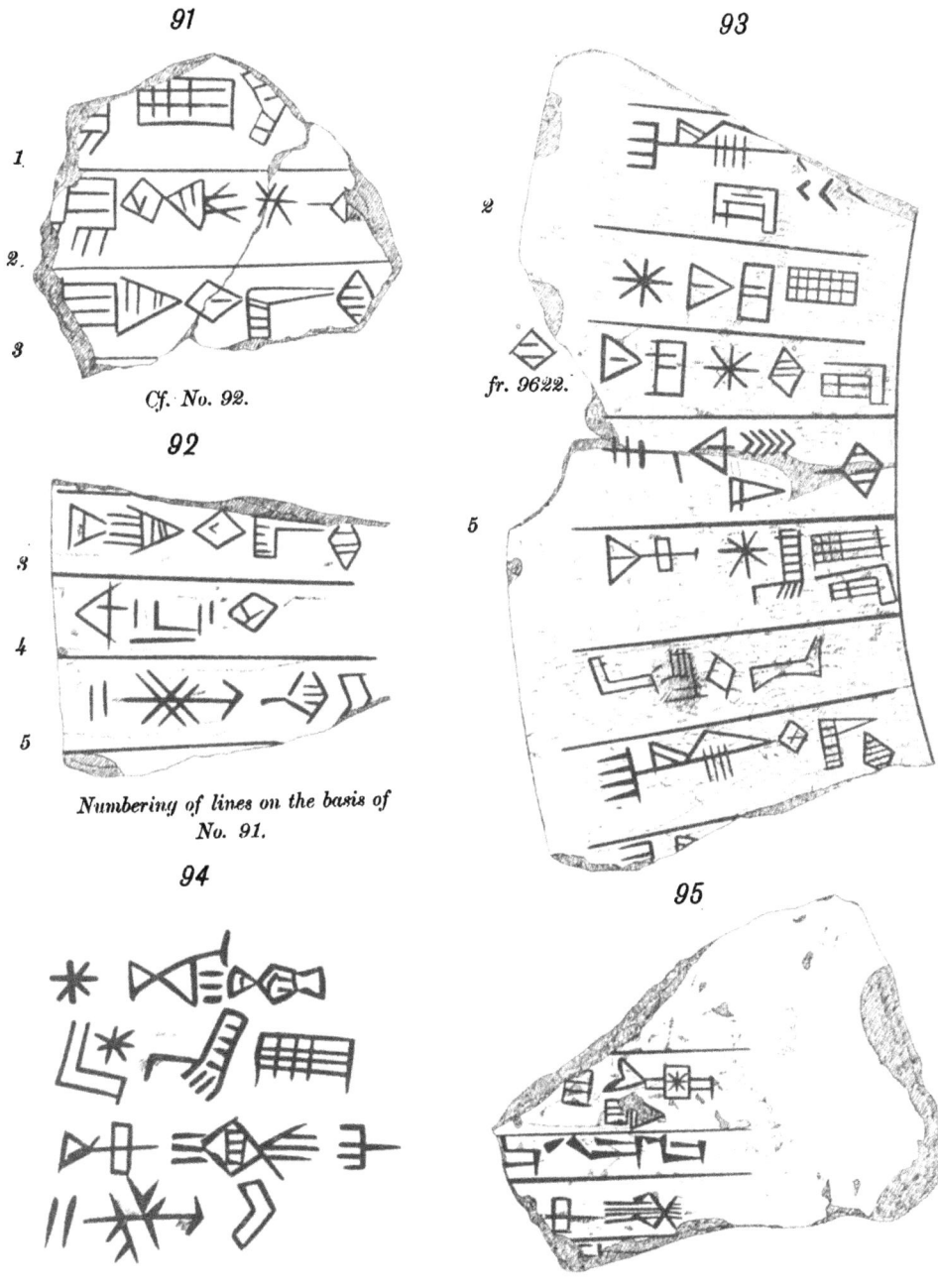

91

1

2

3

Cf. No. 92.

92

3

4

5

Numbering of lines on the basis of
No. 91.

94

93

2

fr. 9622.

5

95

Pl. 44

Trans. Am. Phil. Soc., N. S. XVIII, 8.

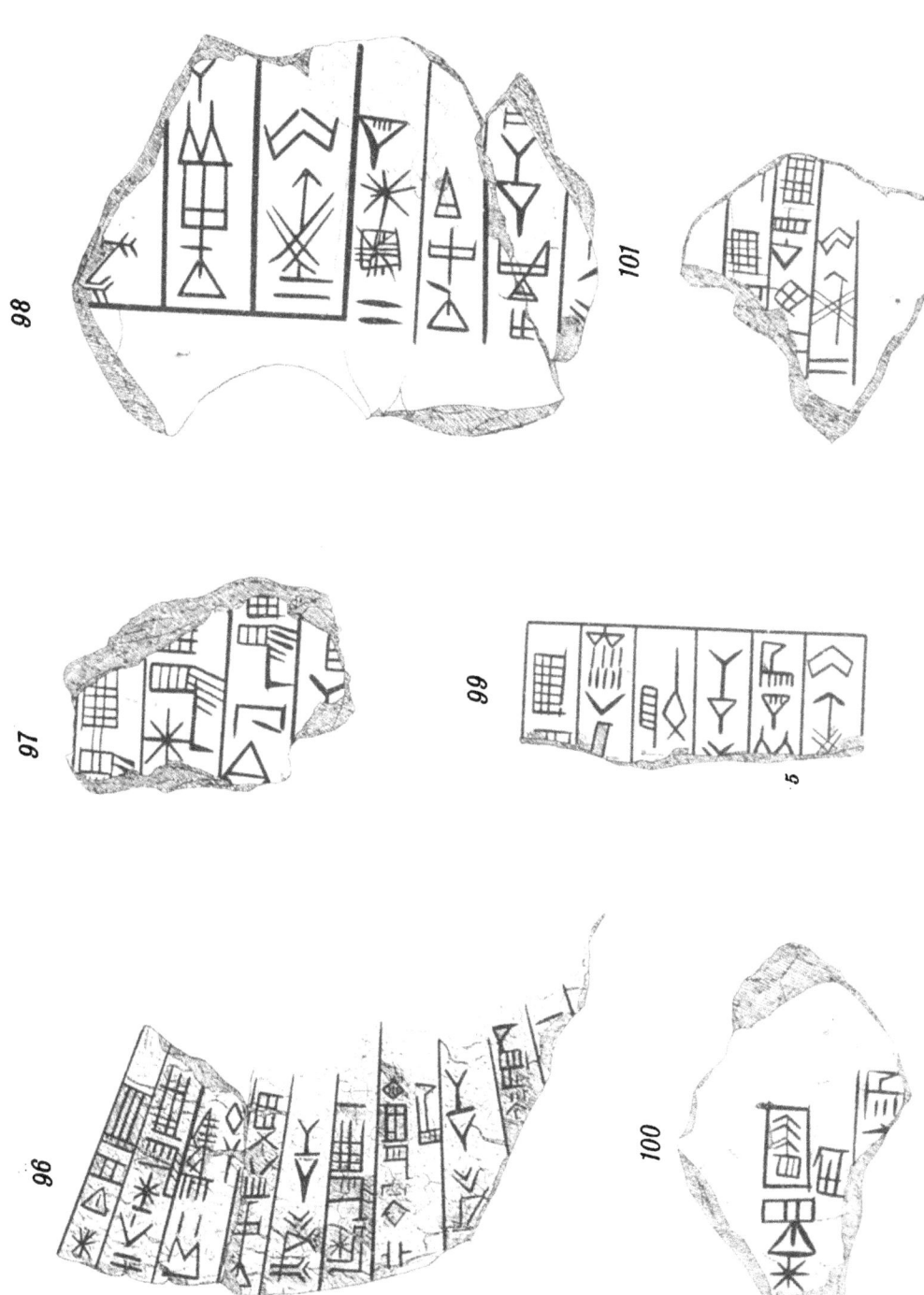

Trans. Am. Phil. Soc., N. S. XVIII, 8.

Pl. 45

103

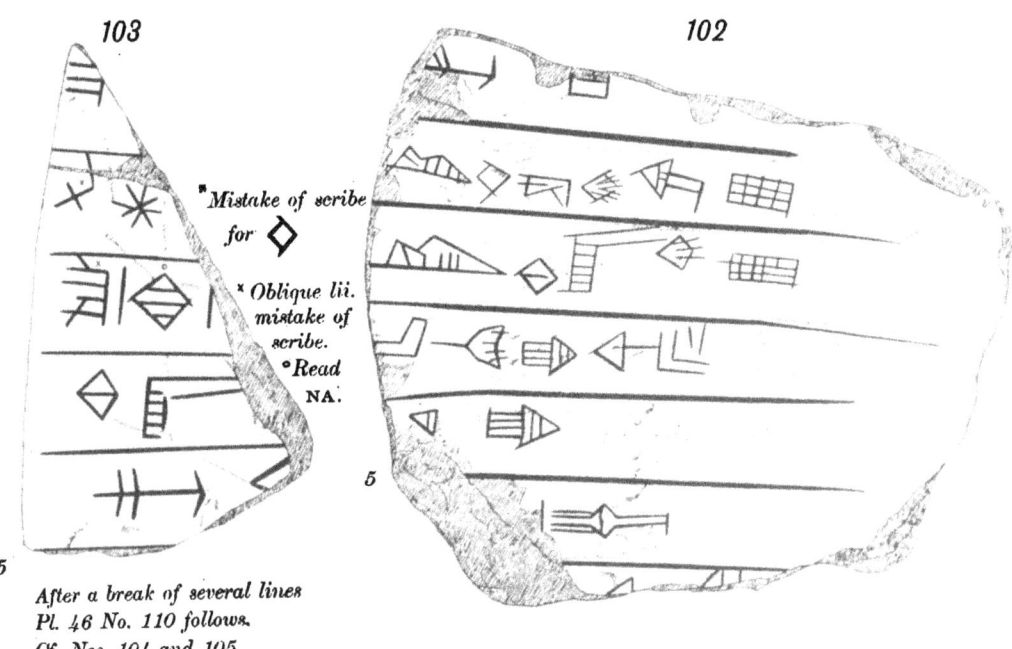

Mistake of scribe
for ◇

ˣ *Oblique lii.*
mistake of
scribe.
° *Read*
NA.

102

After a break of several lines
Pl. 46 No. 110 follows.
Cf. Nos. 104 and 105.

104

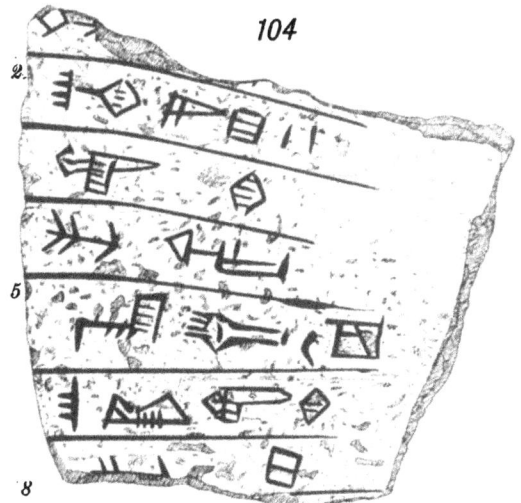

Numbering of lines on the basis of No. 103.
Cf. No. 105.

105

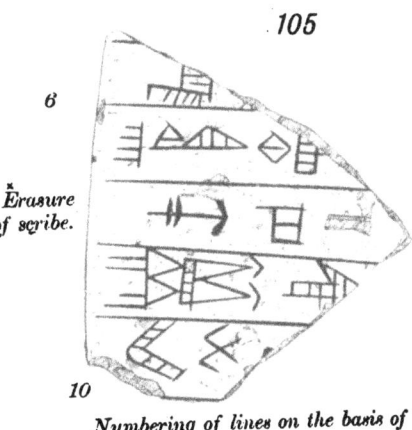

ˣ *Erasure*
of scribe.

Numbering of lines on the basis of
Nos. 103 and 104.

107

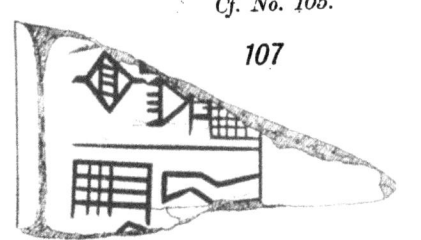

106

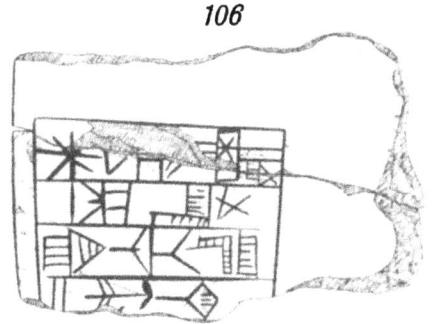

Pl. 46

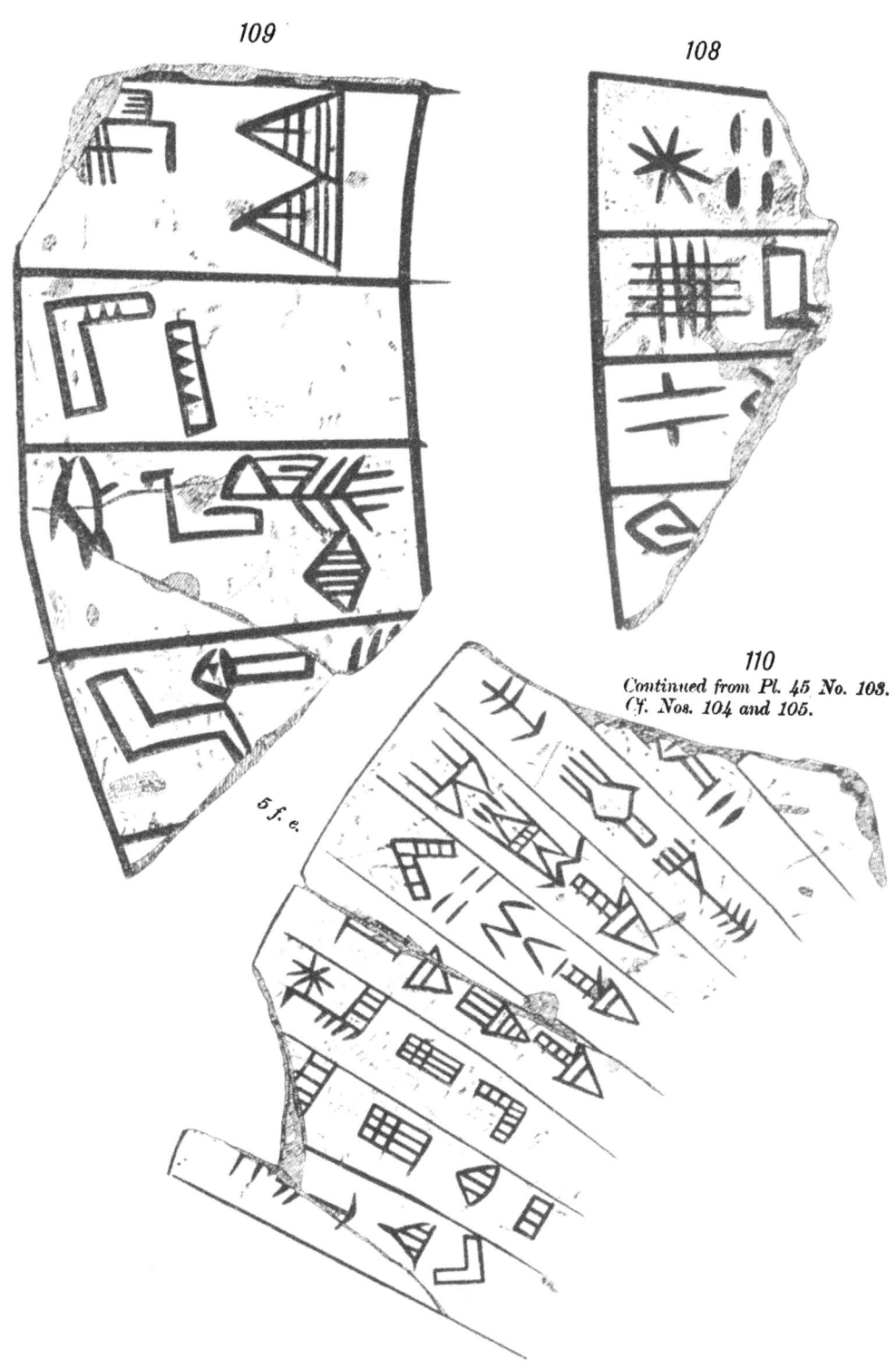

109

108

110
Continued from Pl. 45 No. 103.
Cf. Nos. 104 and 105.

5 f. e.

Pl. 47

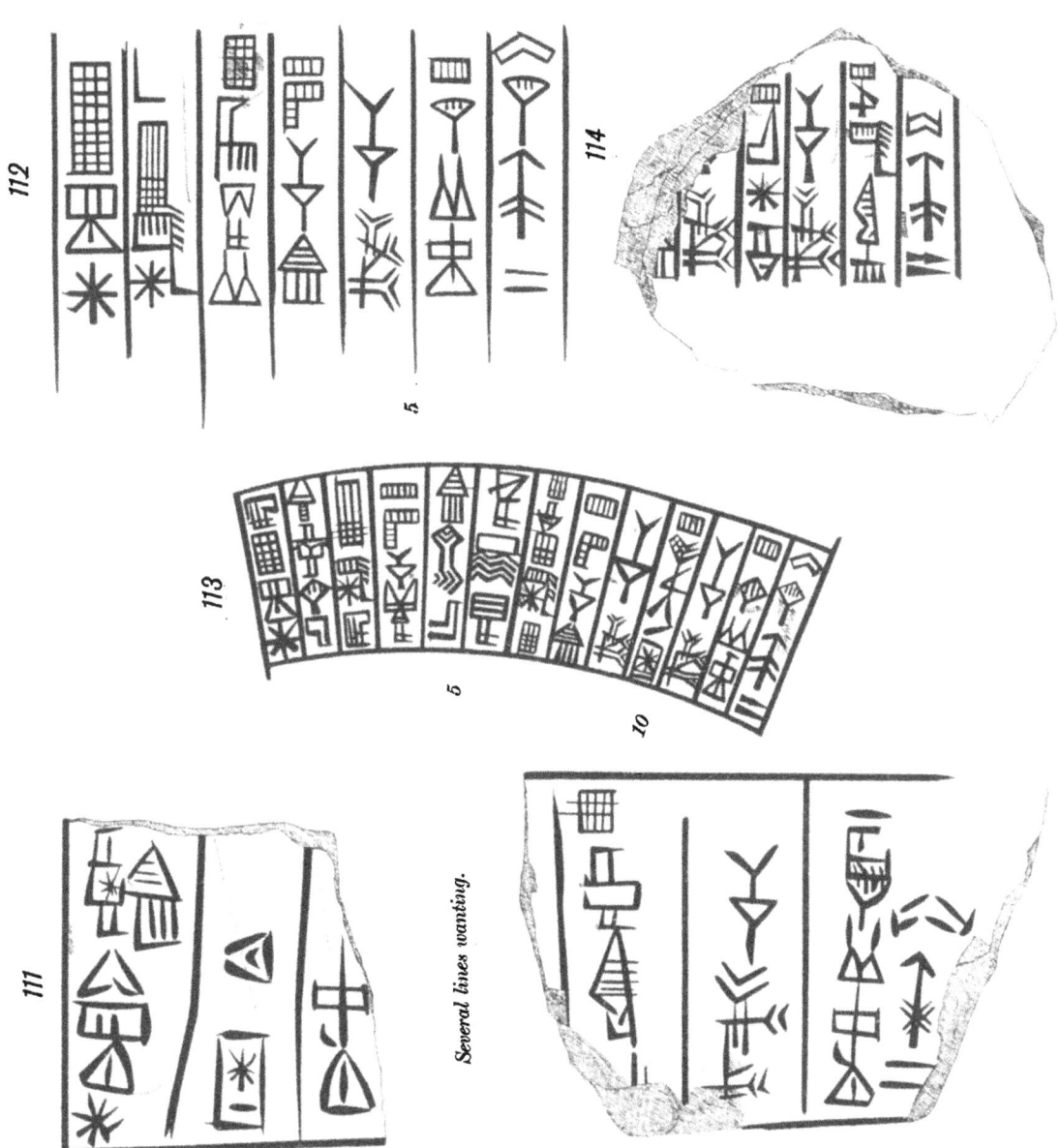

Several lines wanting.

Pl. 48

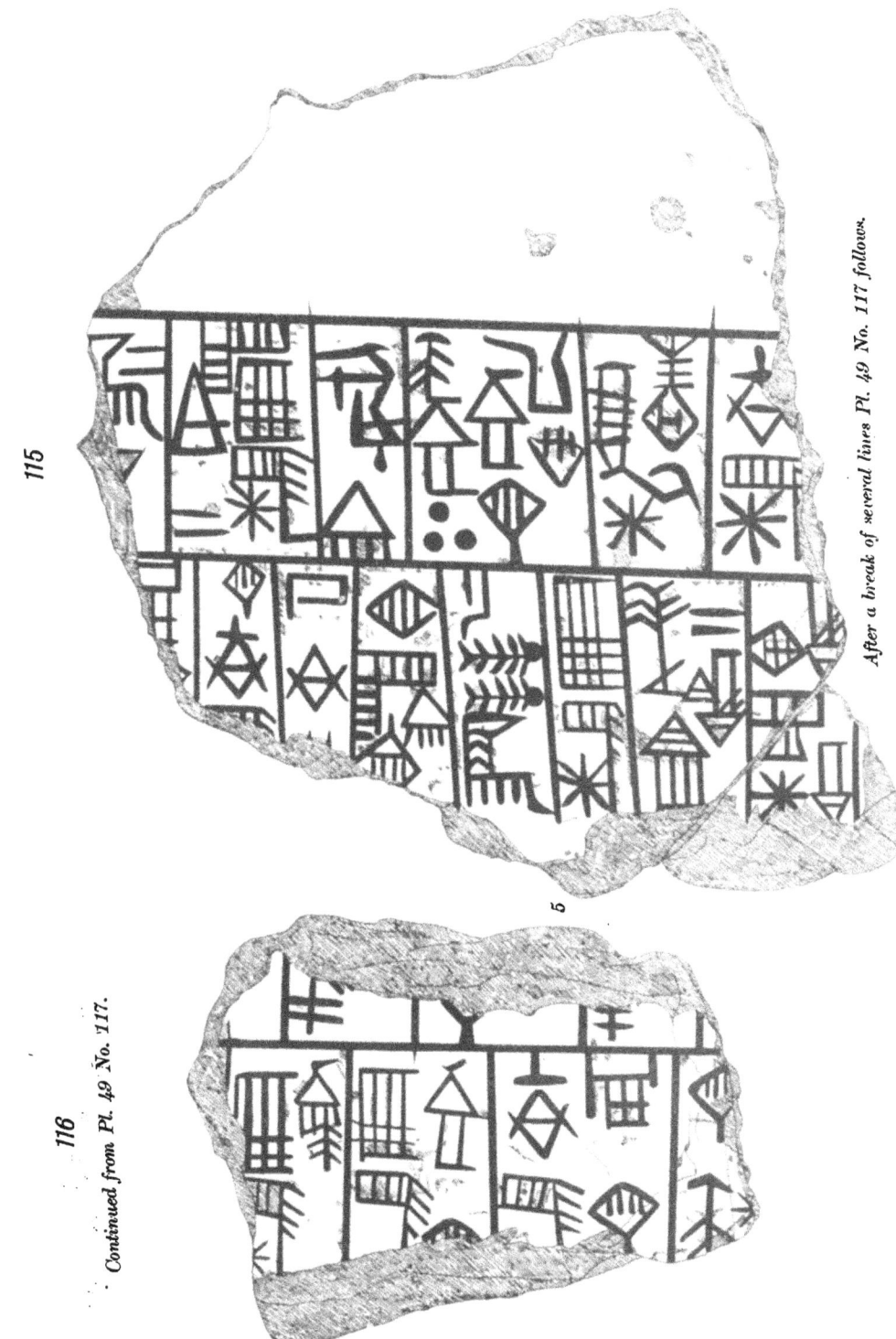

115

After a break of several lines Pl. 49 No. 117 follows.

116

Continued from Pl. 49 No. 117.

Trans. Am. Phil. Soc., N. S. XVIII, 8.

Pl. 49

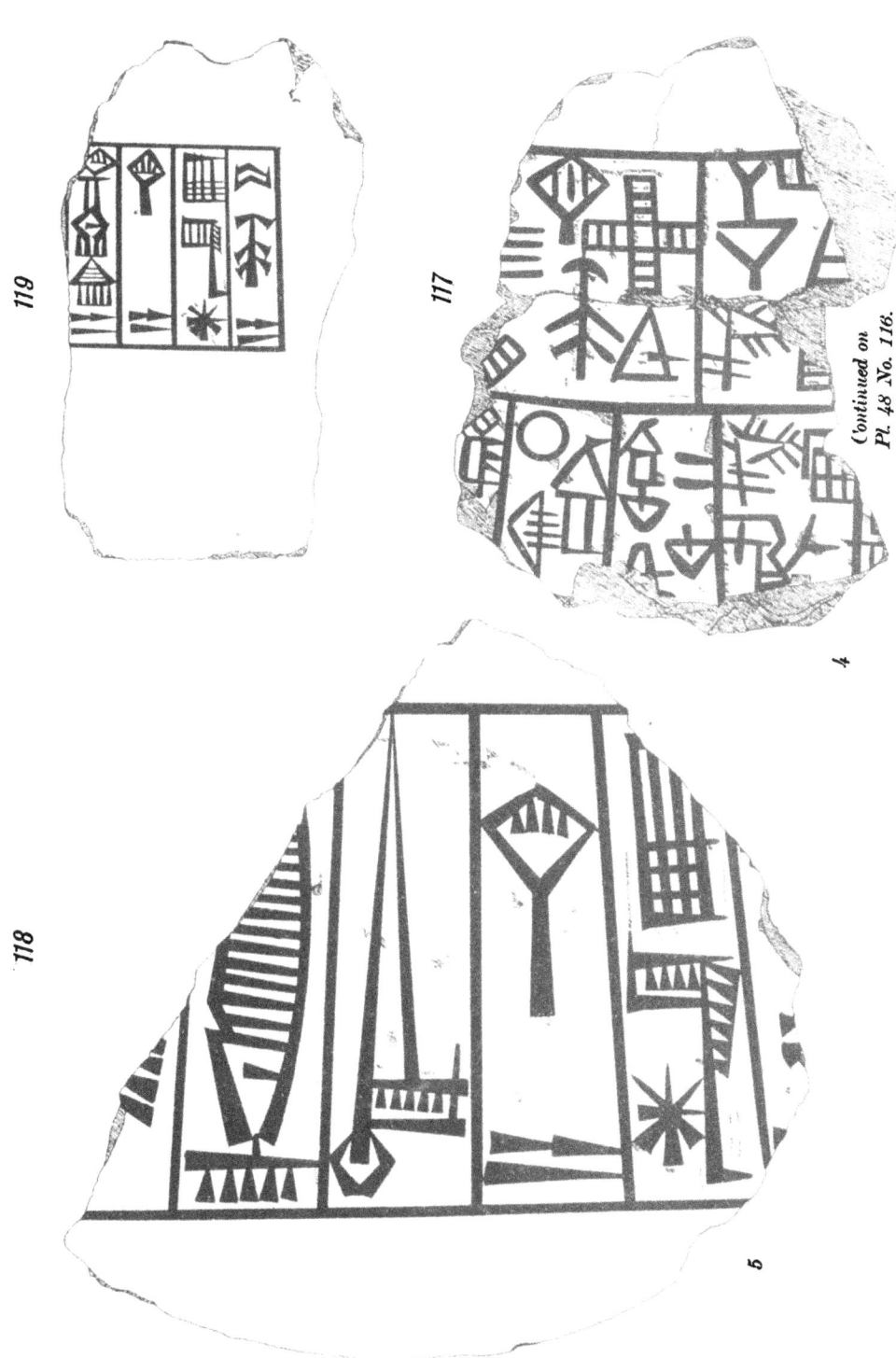

Continued on Pl. 48 No. 116.

Pl. 50

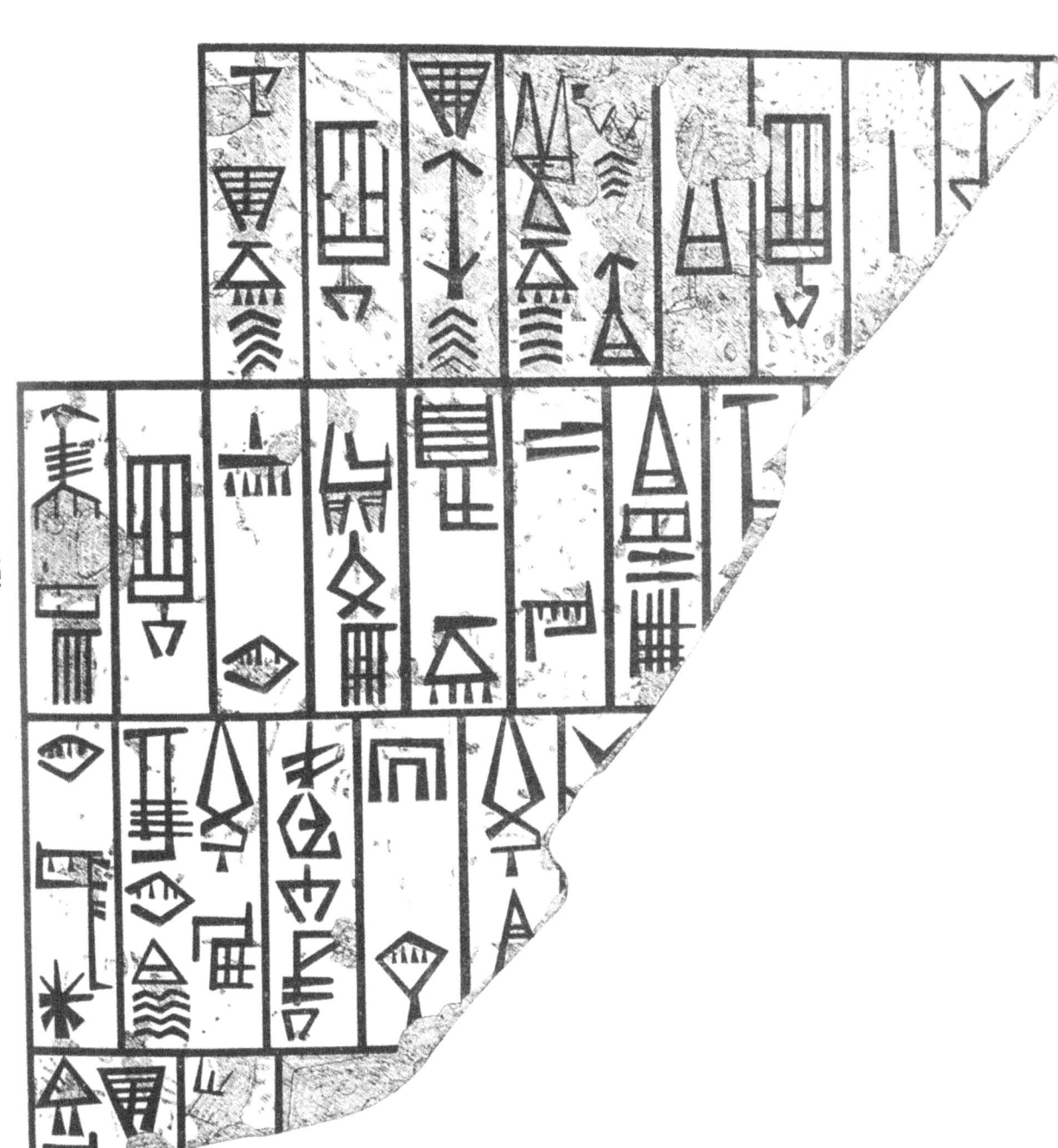

Pl. 51.

121

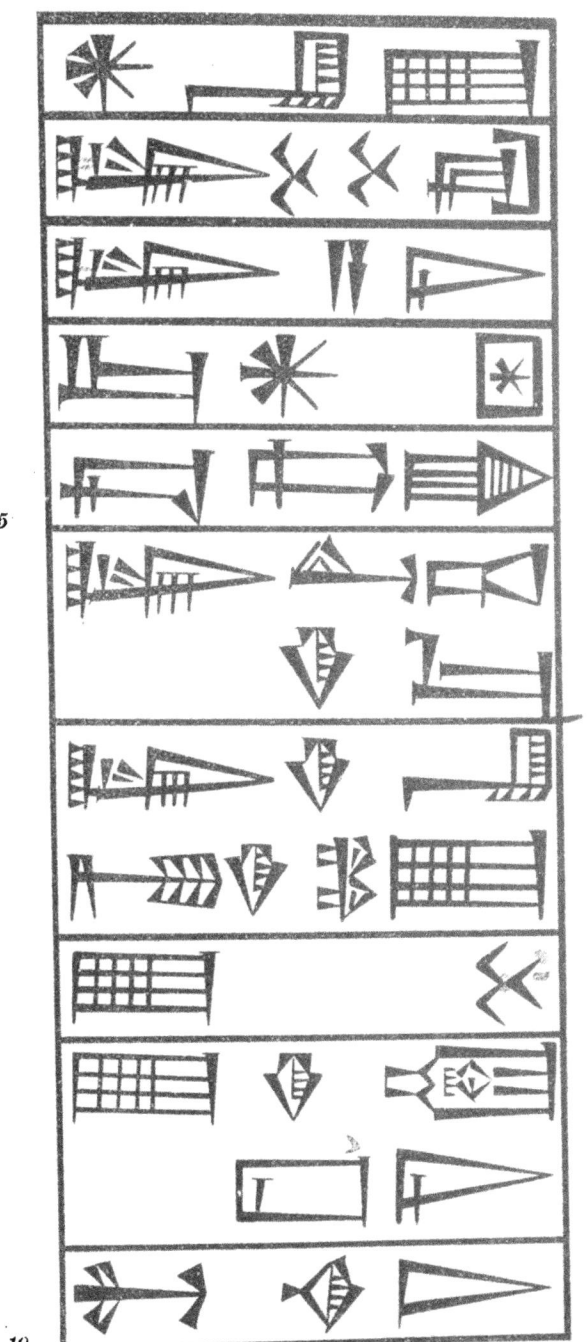

Pl. 52.

122

123

Obverse.

Reverse.

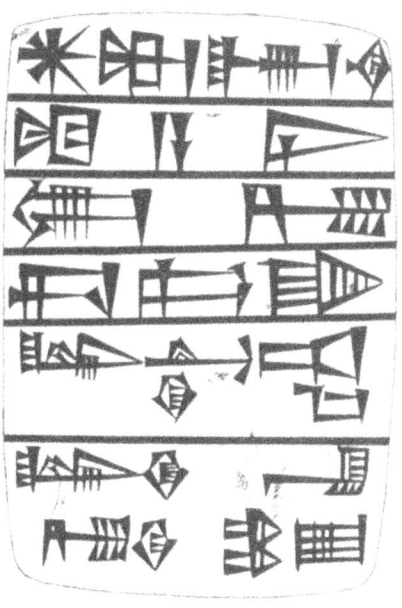 

Pl. 53

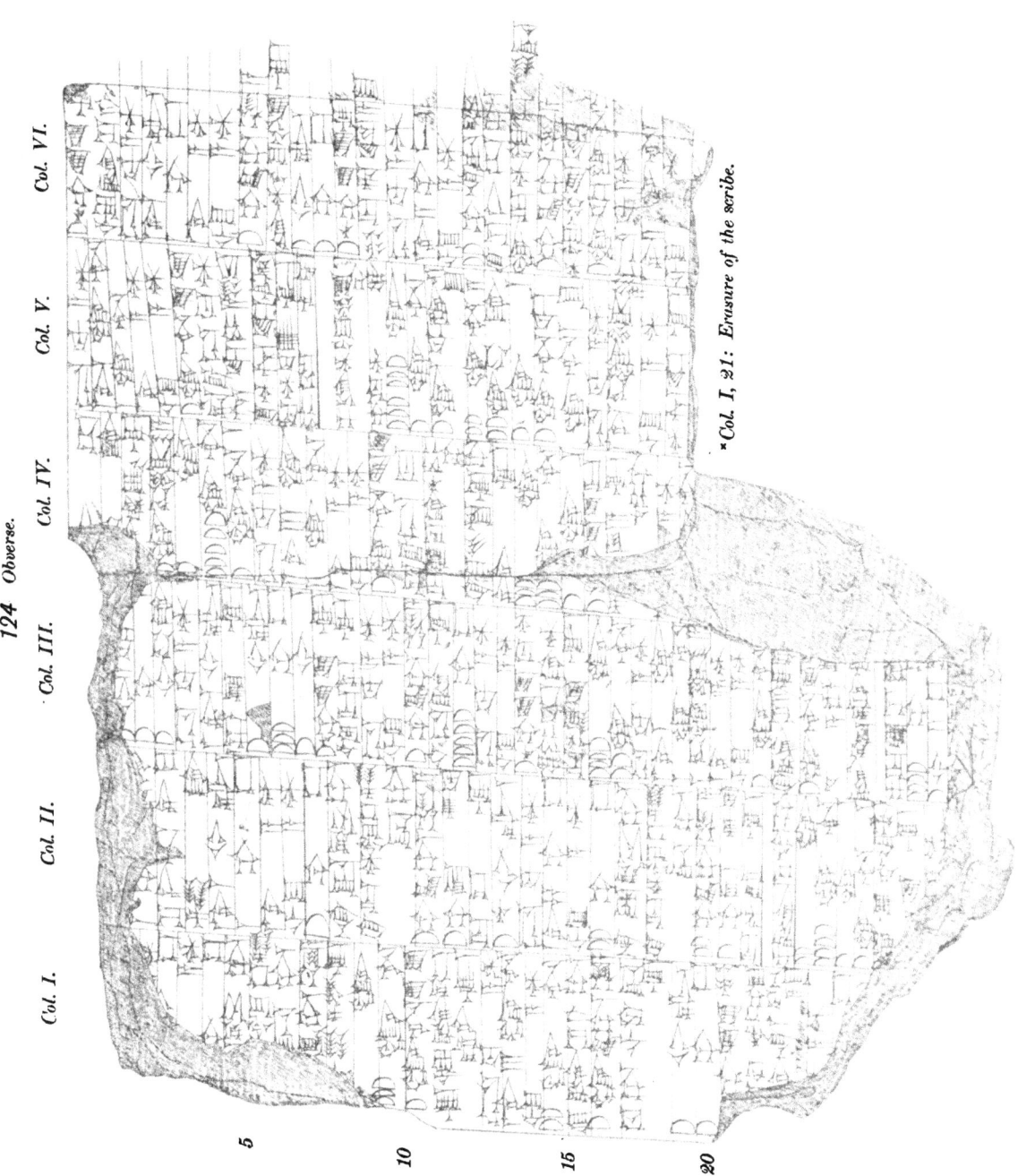

124 Obverse.

Col. I. Col. II. Col. III. Col. IV. Col. V. Col. VI.

*Col. I, 21: Erasure of the scribe.

Pl. 54

Col. I.

Col. II.

Col. III.

Col. IV.

Col. V.

Col. VI.

124 Reverse.

Pl. 55

Reverse.

125

Obverse.

Pl. 56

126

Obverse.

Col. I.   Col. II.   Col. III.   Col. IV.   Col. V.   Col. VI.   Col. VII.

Trans. Am. Phil. Soc., N. S. XVIII, 8,

Pl. 57

126

Reverse.

| Col. VII. | Col. VI. | Col. V. | Col. IV. | Col. III. | Col. II. | Col. I. |

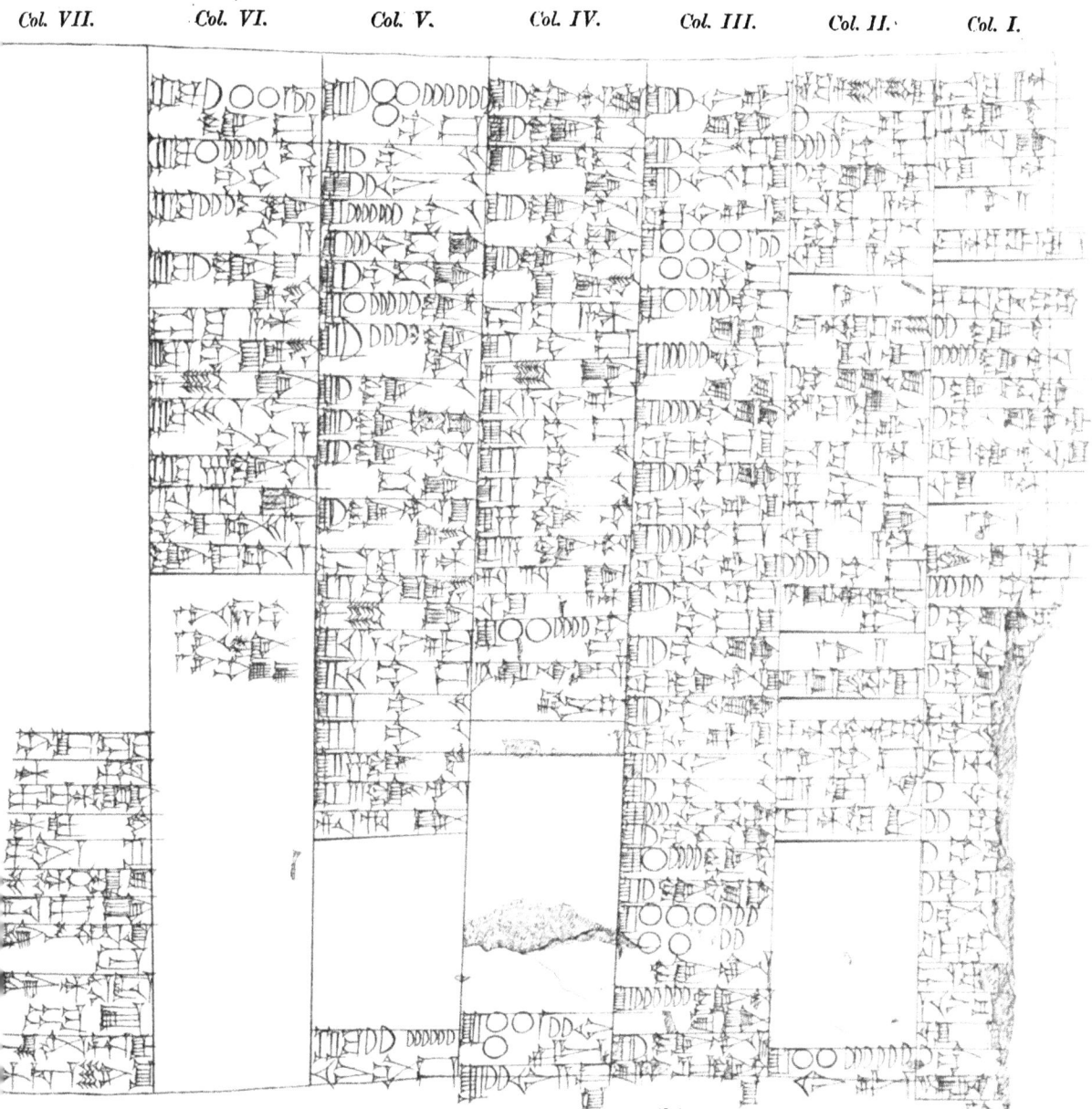

*Col. IV, 11, 12, 6, 19:   Col. V, 8, 10, 20: Erasure of the scribe.

Pl. 58

127

Obverse.

Reverse.

Pl. 59

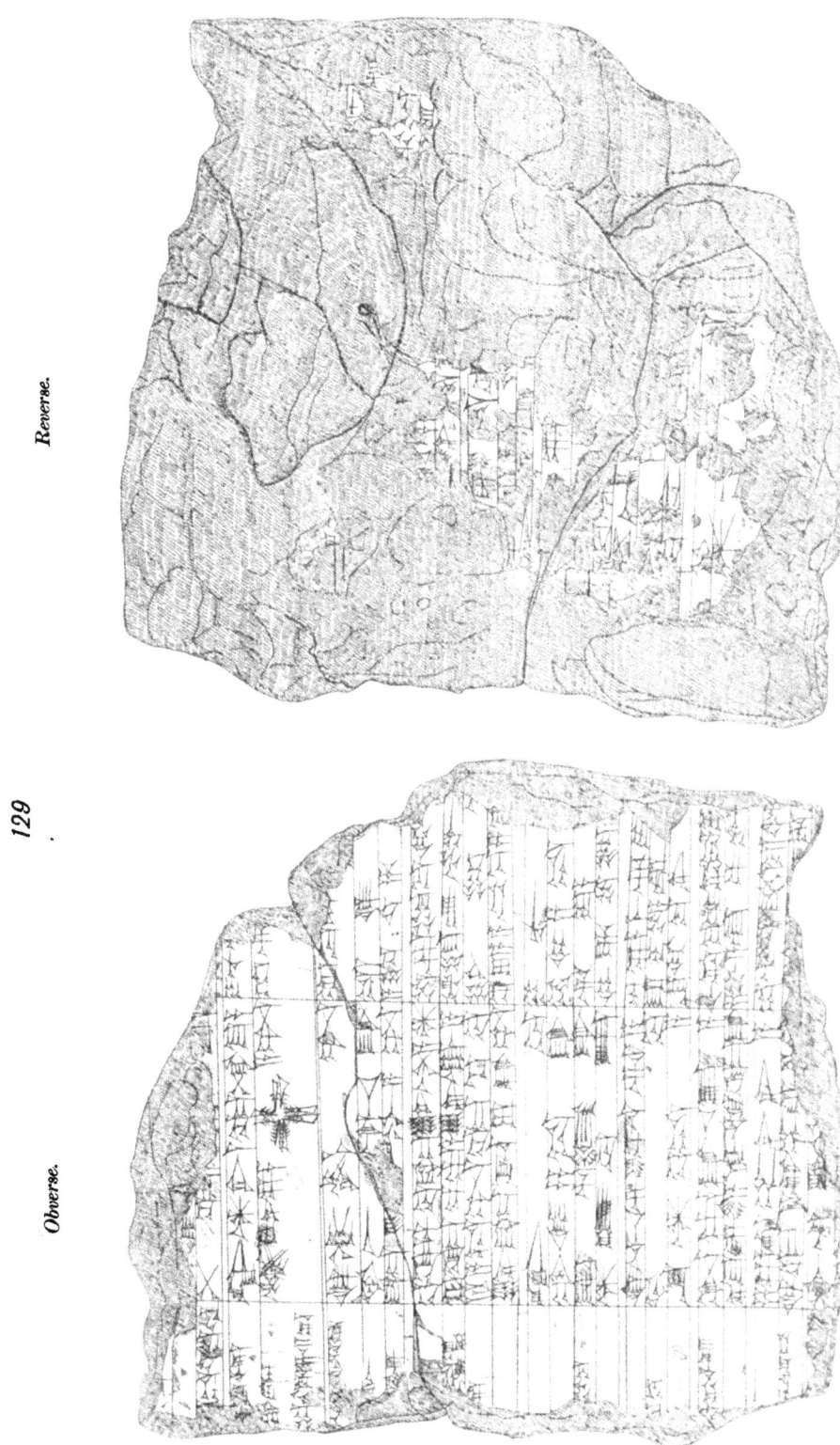

Reverse.

129

Obverse.

Pl. 60

130

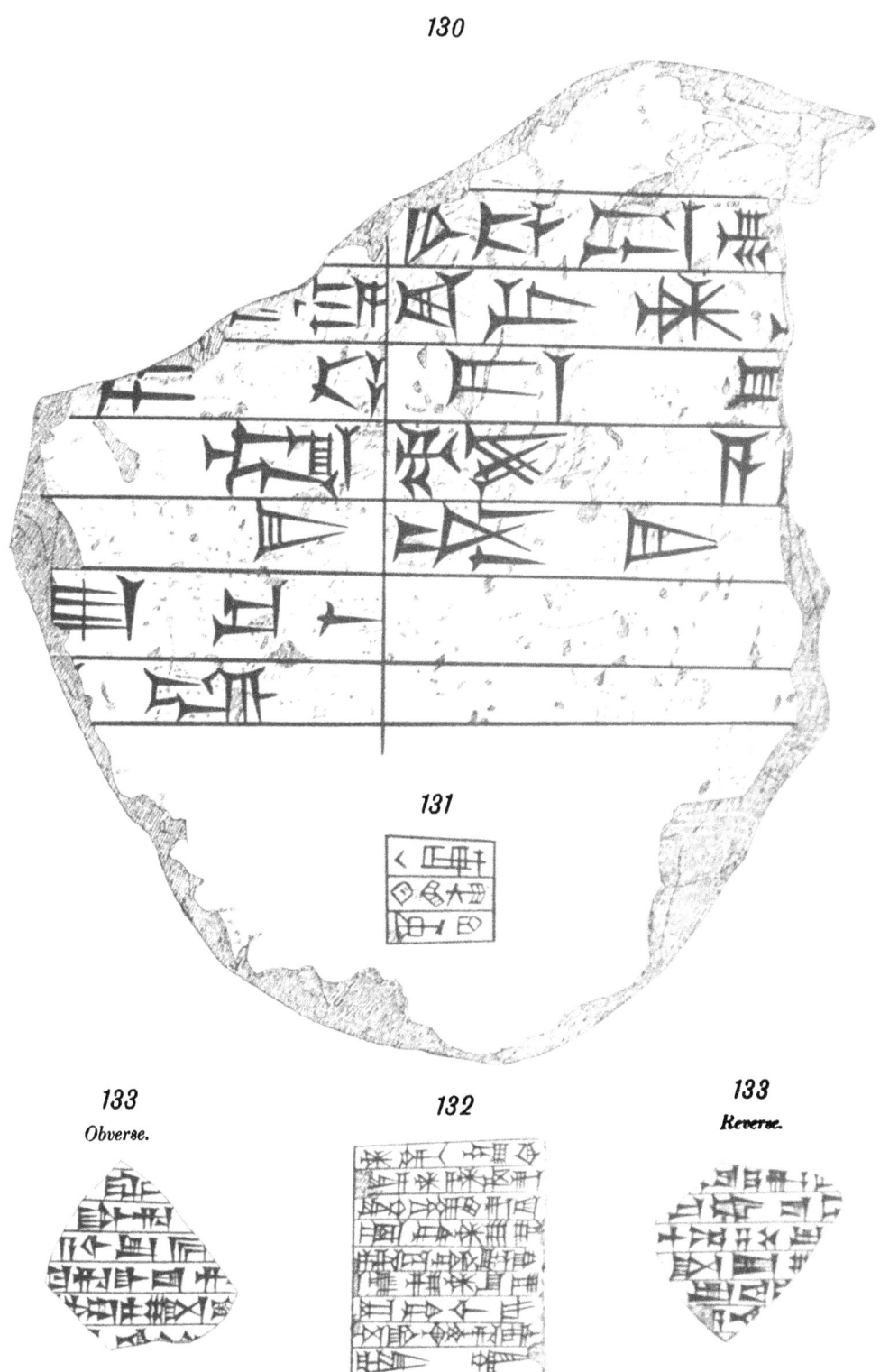

131

133
Obverse.

132

133
Reverse.

Pl. 61

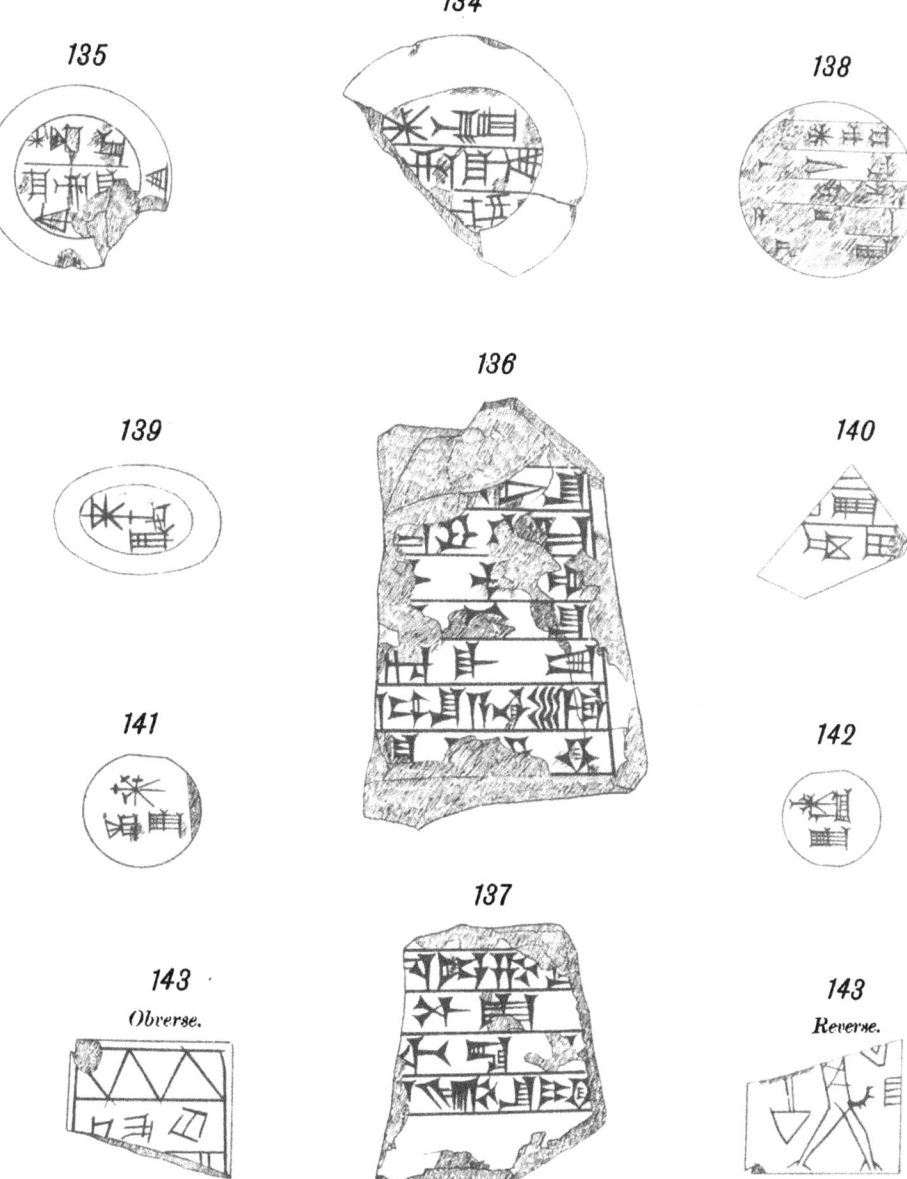

Pl. 62

144

Obverse.

Reverse.

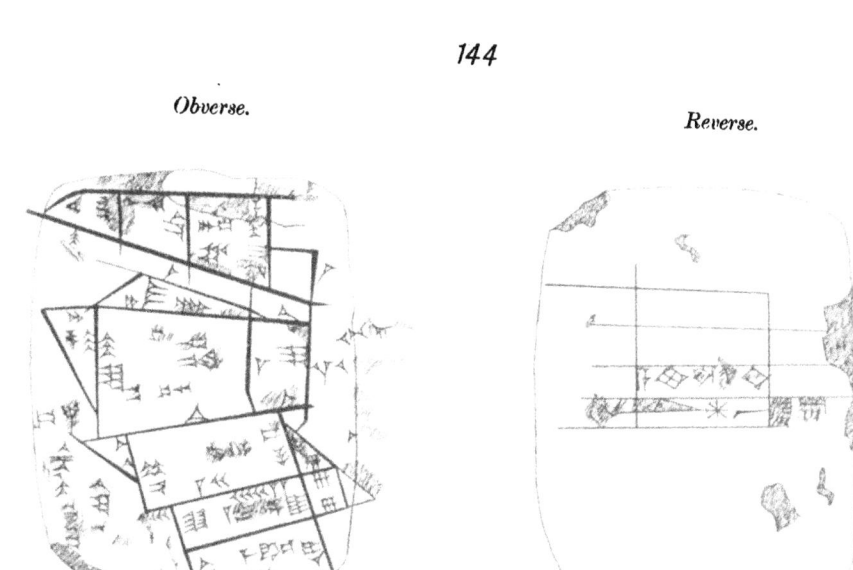

145

Obverse.

Reverse.

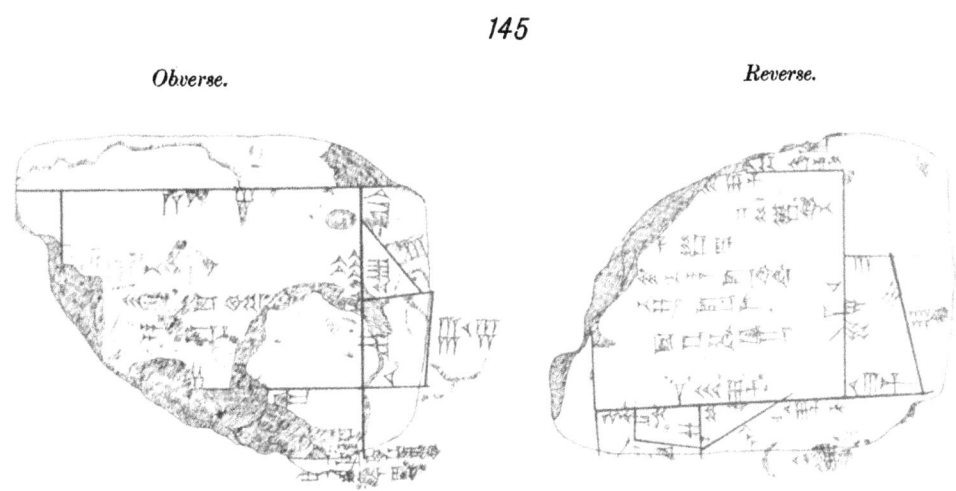

Pl. 63

146

Col. I. Col. II. Col. III. Col. IV.

×Col. III, 17: Read ⪡ the rest is erasure of the scribe.

Col. III, 38: Read ⪦ the rest is erasure of the scribe.

Pl. 64

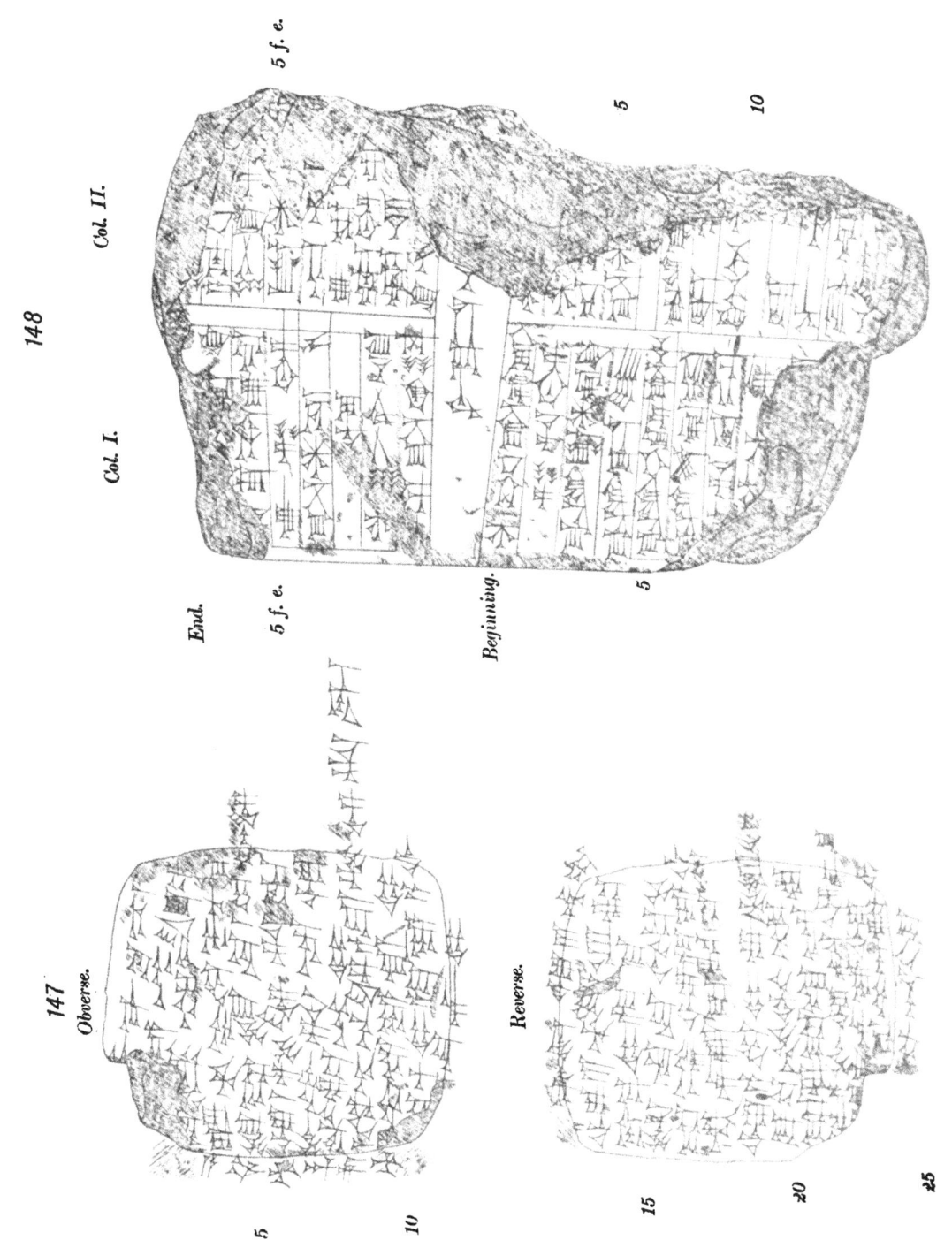

148

Col. II.

Col. I.

5 f. e.

End.

5 f. e.

Beginning.

5

10

5

147

Obverse.

Reverse.

5

10

15

20

25

Trans. Am. Phil. Soc., N. S. XVIII, 3.

Pl. 65

149

Col. I.

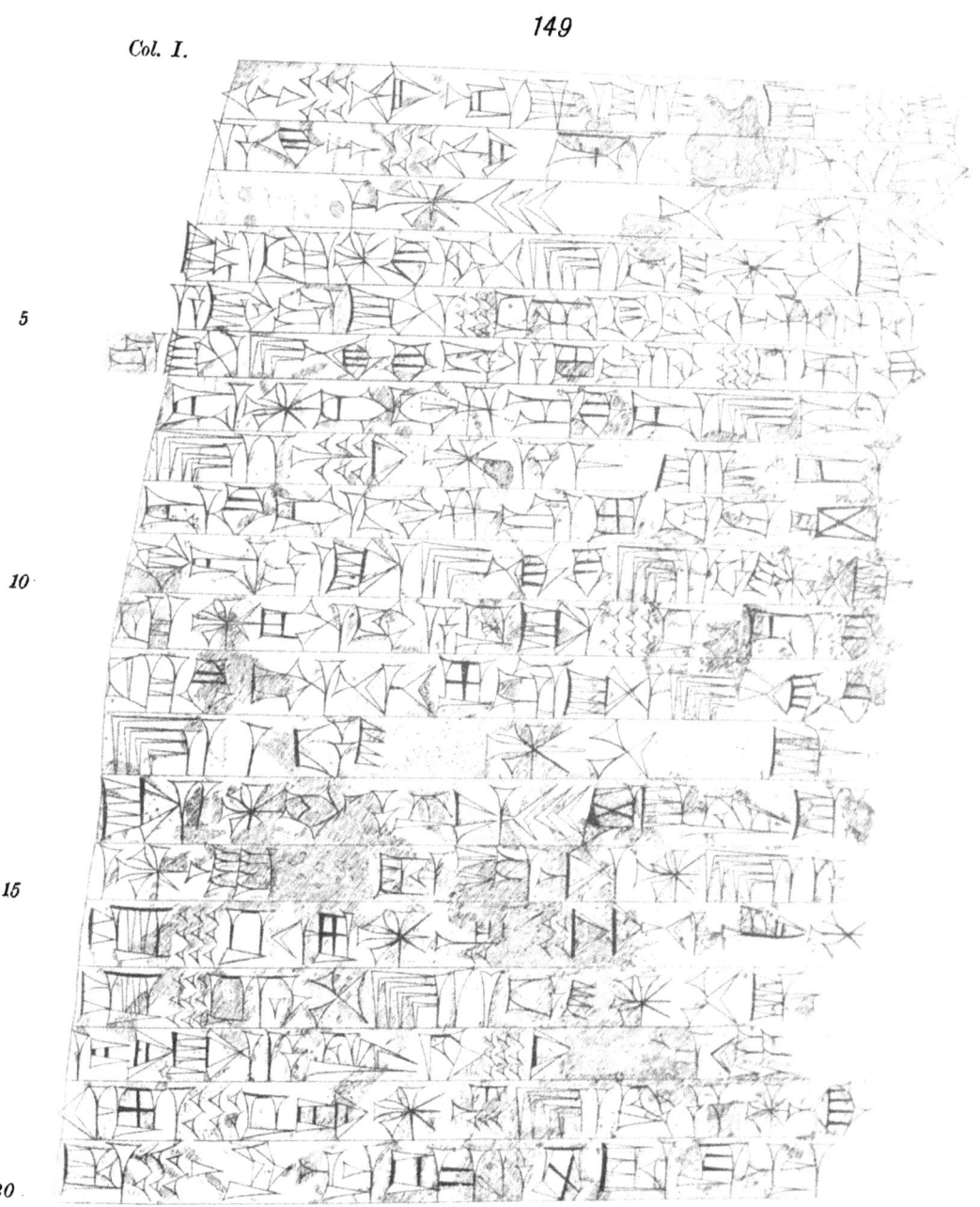

5

10

15

20

Trans. Am. Phil. Soc., N. S. XVIII, 8.

Pl. 66

149
Continued

Col. I.

21

Col. II.

5

10

15

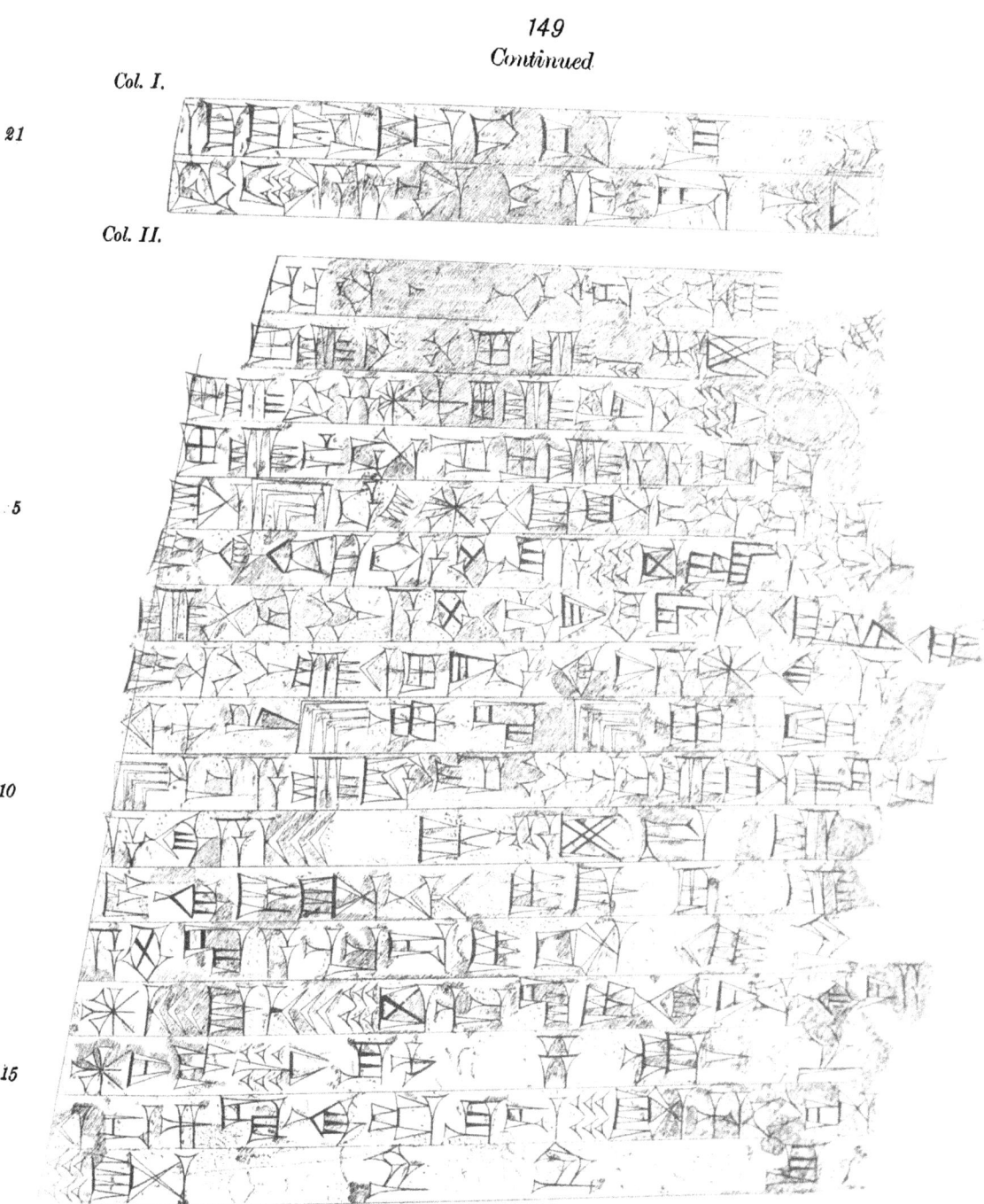

Pl. 67

149
Continued

Col. II.

20

Col. III.

5

10

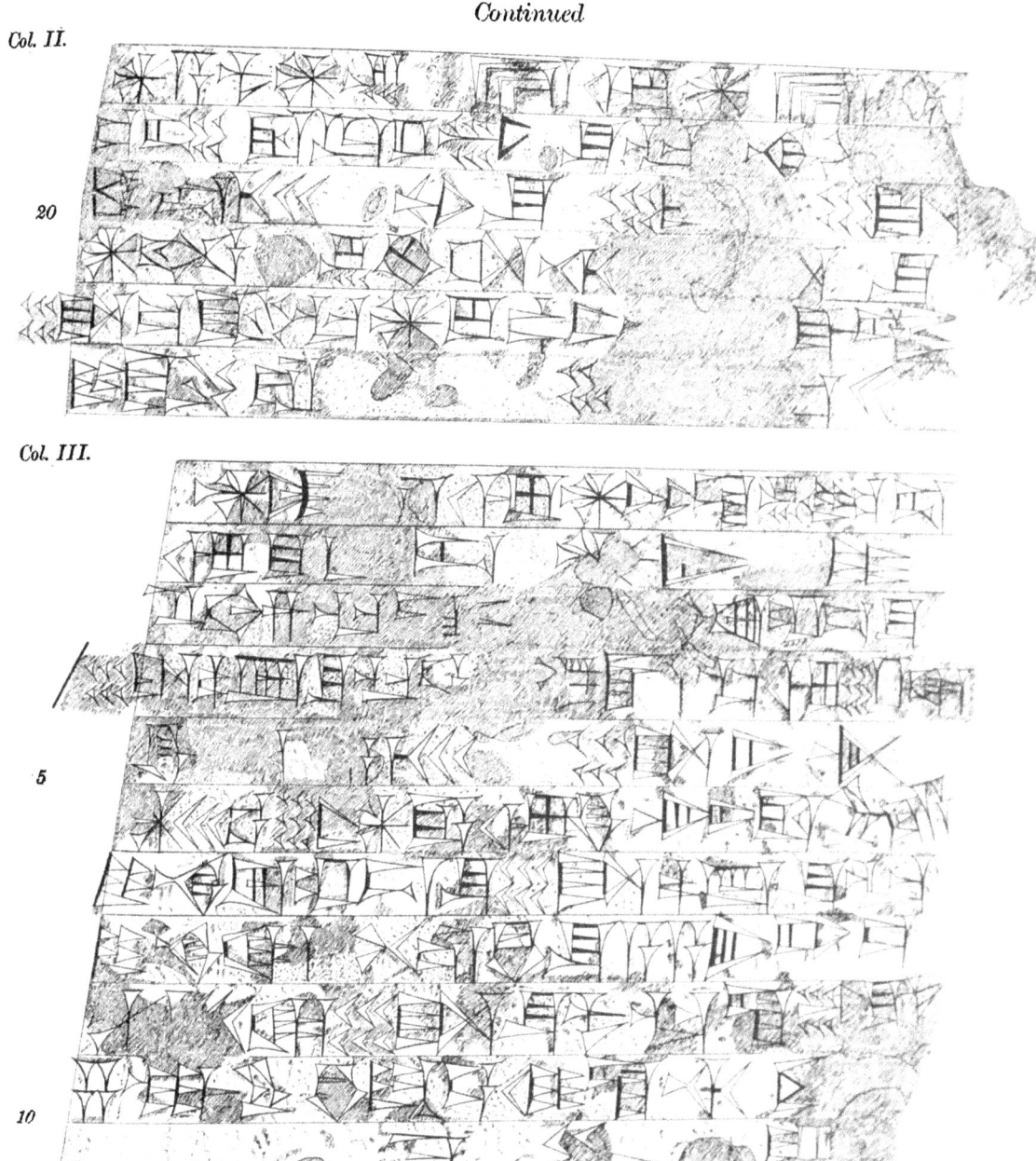

Pl. 68

150.

Col. I.

Col. II.

Pl. 69

151

L. 3: Erasure of the scribe.

Pl. 70

152

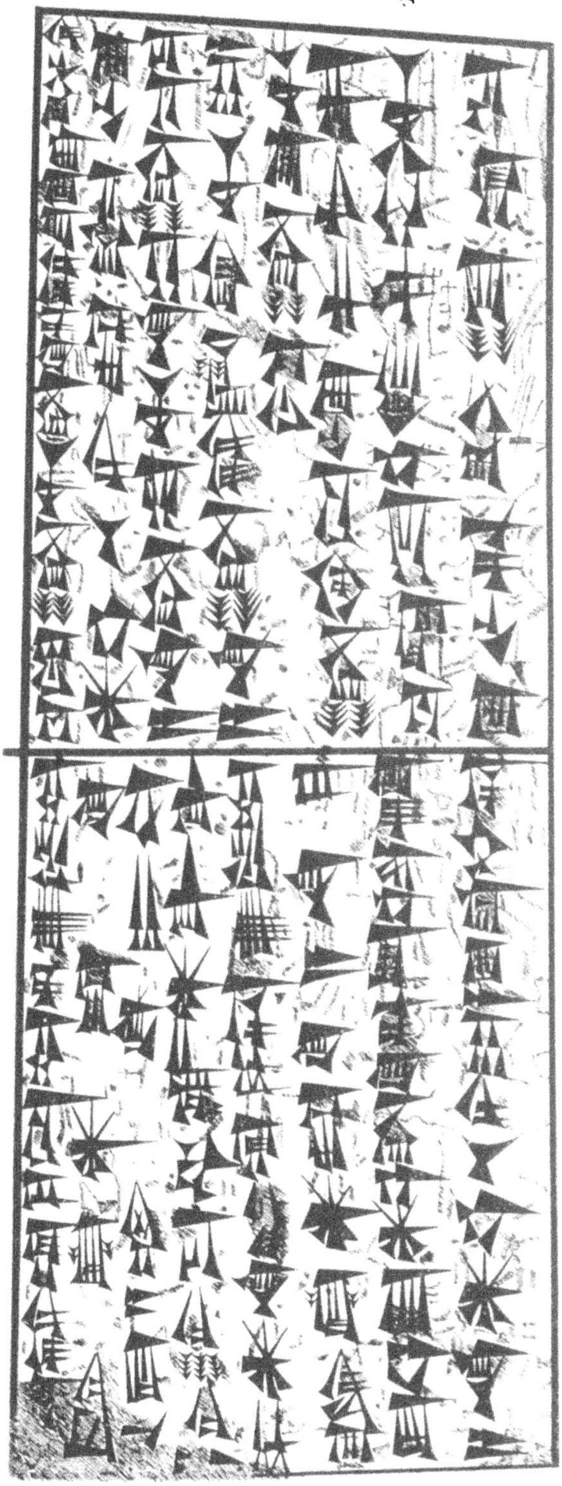

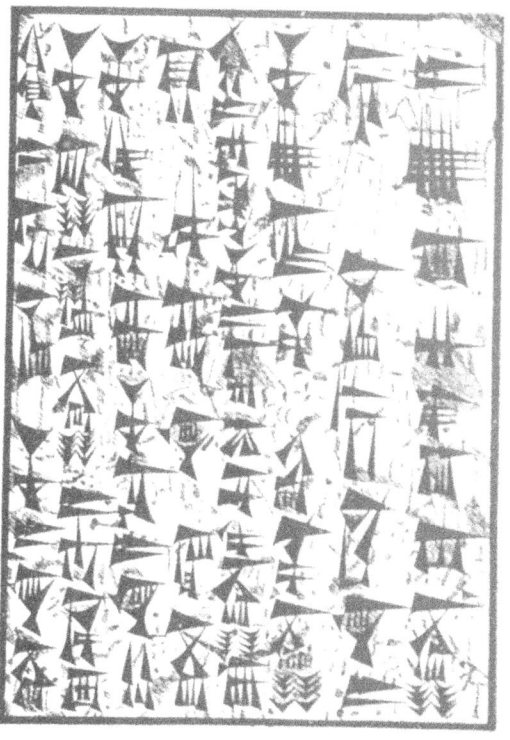

87

88

VOTIVE TABLETS IN LIMESTONE, INCISED.
Nippur.

Trans. Am. Phil. Soc., N. S. XVIII, 8.

PL. XVII

89

MARBLE BLOCK OF LUGALKIGUBNIDUDU,
Nippur,

VASE FRAGMENTS OF LUGALKIGUBNIDUDU.
Nippur.

Trans. Am. Phil. Soc., N. S. XVIII. 8.

VASE FRAGMENTS OF LUGALZAGGISI.
Nippur.

Trans. Am. Phil. Soc., N. S. XVIII, 3.

PL. XX

62

VASE OF ALUSHARSHID (URU-MU-USH).
Nippur.

68

BRICK OF SARGON I.
Nippur.

Trans. Am. Phil. Soc., N. S. XVIII, 8.

PL. XXII

64

INSCRIBED BAS-RELIEF OF NARAM-SIN.
Diarbekir.

Trans. Am. Phil. Soc., N. S. XVIII, 8.

PL. XXIII

65

BRICK OF UR-NINIB—Nippur.

Inscription begins at bottom.

Trans. Am. Phil. Soc., N. S. XVIII, 8.

PL. XXIV

66

67

68

66, 67. CLAY TABLET (OBVERSE AND REVERSE).—Tell el-Hesy.
68. Fragm. of a barrel-cylinder of Mardukshabikzerim.—Place unknown.

69

70

69. Fragm. of a Boundary Stone.   70. Inscribed Pebble.
Nippur.

71

BAS-RELIEF IN CLAY WITH AN ARAMAIC INSCRIPTION.
Nippur.

Trans. Am. Phil. Soc., N. S. XVIII, 8.

PL. XXVII

72

TERRA COTTA VASE WITH ROPE PATTERN, C. 4000 B. C.—Nippur.

Height, 68.5 cm.; diameter at the top, 58 cm.

Found in an upright position 5.49 m. below the eastern foundation of Ur-Gur's Ziggurrat, and 3.05 m. below a pavement which consists entirely of burned bricks of Sargon I and Narâm-Sin. It stood 7 m. south-east from an altar, the top of which was c. 2.40 m. higher than that of the vase.

78

ARCH OF BURNED BRICK LAID IN CLAY MORTAR, C. 4000 B. C.—Nippur.

71 cm. high, 51 cm. span, 88 cm. rise.

At the orifice of an open drain passing under the eastern corner of Ur-Gur's Ziggurrat, c. 7 m. below the foundation of the same, and 4. 57 m. below a pavement which consists entirely of burned bricks of Sargon I and Narâm-Sin. View taken from inside the drain. Front of arch opened to let light pass through.

74

NORTH-WESTERN FACADE OF THE FIRST STAGE OF UR-GUR'S ZIGGURRAT.
Nippur.

75

GENERAL VIEW OF THE EXCAVATIONS AT THE TEMPLE OF BEL.—SOUTH-EAST SIDE.

1, 6 (8), 7 (9)—Three stages of the Ziggurrat. 1—East corner of Ur-Gur's Ziggurrat. 2—Excavated rooms on the south-east side of the temple and separated from the latter by a street. 3—Causeway built by Ur-Gur, leading to the entrance of the Ziggurrat. 4—Deep trench extending from the great wall of the temple enclosure to the facade of Ur-Gur's Ziggurrat, 5—Modern building erected by Mr. Haynes in 1894, after an unsuccessful attempt by the Arabs to take his life.